KU-315-711

Corporate Scriptwriting

A Professional's Guide

Ray DiZazzo

Focal Press
Boston   London

Leeds College LIBRARY 5 JUl 1994 of Art and Design

£32.50

791. 81

R 11223 Y

Focal Press is an imprint of Butterworth–Heinemann.

Copyright © 1992 by Butterworth–Heinemann, a division of Reed Publishing (USA) Inc.
All rights reserved.

No part of this publication may be reproduced, stored in a retrieval system, or transmitted, in
any form or by any means, electronic, mechanical, photocopying, recording, or otherwise,
without the prior written permission of the publisher.

 Recognizing the importance of preserving what has been written, it is the policy of Butterworth–
Heinemann to have the books it publishes printed on acid-free paper, and we exert our best
efforts to that end.

Figures 3.1, 3.2, and 20.2 reproduced with permission of Alan C. Ross.
Figures 4.1, 8.1–8.7, and scripts reproduced with permission of VisNet™ West,
GTE Service Corporation. Growth and survival tips reproduced with consent.

Library of Congress Cataloging-in-Publication Data

DiZazzo, Raymond.
    Corporate scriptwriting : a professional's guide / by Ray DiZazzo.
        p.   cm.
    Includes bibliographical references and index.
    ISBN 0-240-80115-6 (pbk. : alk. paper)
    1. Television authorship.   2. Industrial television—Authorship.
    I. Title.
    PN1992.7.D59   1992
    808'.066791—dc20   91-24335
                                                              CIP

British Library Cataloguing in Publication Data

DiZazzo, Ray
    Corporate scriptwriting : a professional's guide.
    I. Title
    808.23

ISBN 0-240-80115-6

Butterworth–Heinemann
80 Montvale Avenue
Stoneham, MA 02180

10 9 8 7 6 5 4 3 2 1

Printed in the United States of America

For Patti, Sunday, Sean, Gus, Emily, Guy, Dean,
Mac, Cedric, Jimmy, Robert and Allan
— my early supporters.

Contents

Preface

I often remember two concepts from books read in my early days as a student of corporate scriptwriting. The first was based on the idea that the writer was being hired to create and convey a series of script pictures. Conveying these pictures meant first visualizing and then carefully writing down every detail of that visualization—including each camera angle, zoom, and camera movement—into the scene descriptions.

The second concept held that a script should always be written starting with the visuals. The argument for this was that films and videotapes are visual media and that to create a visual program starting with printed words was like trying to build a bridge starting from the top down.

Both concepts seemed logical enough then, and they still do today, except when they bring to mind another recollection from those early days.

After I had managed to get one of my first scriptwriting jobs, I was sitting in a small conference room across from the first producer-director I had ever met. The purpose for our meeting was to give me basic instructions on one of several scripts I had (tentatively, at this point) been contracted to write for him.

"Oh, and by the way," he said, shortly after our conversation began, "on each of the opening segments, don't worry about the visual side of the script at all. Just write 'effects' in on the left side and write the narration in the sound column. We'll be creating the visuals with an art system based on a music track and your copy."

In another meeting with a different producer, the subject of how to handle scene descriptions came up. Before I had the chance to assure him that I had read up on this and would include copious detail in every scene, he said, "No camera this or that. No 'close ups' or 'trucks' or 'dollys' or 'high angles' or any of that stuff. Just tell me the story. I don't want to have to white out the left side of the page and start all over before I direct these shows."

Thus I was faced with my first major scriptwriting contradiction: just what *was* the right way to write a script? Should I write according to what I had read and heard or according to the producers, the people who handed out the checks?

For me, that was a very basic and disturbing question that bothered me for some time. Although I managed to find my own answer to it over a period of years, one of my main goals in writing this book is to save you the anxiety of that process. My hope is that these pages will answer that and many other questions in your mind at the outset of your scriptwriting career. My second major goal is to provide you with a real-world collection of basic skills with which to launch that career.

After getting well into the manuscript, I found that accomplishing these two goals required exploring the subject of writing scripts from what at first seemed like an odd perspective: scriptwriting as a business rather than as an art. Although this business idea seemed strange at first, I soon realized it was a perfectly logical approach. This is because a successful corporate scriptwriter must have two distinct and equally important abilities:

1. the ability to create and express visual concepts with clarity and sound structure on the written page
2. the ability to handle oneself with grace, ease and flexibility in a business environment

The reason for these two requirements should be obvious: corporate scriptwriting is just as much a business activity as an artistic one. A corporate script is not a work of art but rather art for work. Although writing takes a great deal of artistic talent, creativity, and skill, it is nonetheless being paid for by a business manager to communicate information to a business audience in a business environment.

It follows, then, that corporate scriptwriters must be much more than artistic and creative. They must be as comfortable having a formal lunch with middle managers or even a CEO as they are with fellow writers at a local coffee shop. They should wear business dress, have business cards, and know as much about business politics as they do about the principles of good dialogue.

With these ideas in mind, I have arranged *Corporate Scriptwriting* into four parts. The first, Chapters 1 and 2, deals with defining the business environment and establishing the purpose the script serves in it. The second part is made up of Chapters 3 through 12. This material deals with what I call the writer's basics, the general skills a writer must know and apply on every assignment. Chapters 13 through 18, which make up the third section of this book, deal with the writer's work. They contain complete writing examples that are documented results of those basic skills: a needs assessment, a content outline, a treatment, and several scripts, along with comments on the creative and business considerations that helped guide their development. Finally, Part IV, comprised of Chapters 19 and 20, offers some tips on editing and revising your work and getting started in the field.

As you move through these pages, you will also find three other elements: "You Script It" quizzes, growth and survival tips, and case studies. These are meant to test your skills occasionally and also to pass on valuable advice from professional scriptwriters, producers, directors, and clients who are currently active in the field.

Answers to the "You Script It" quizzes are in Appendix 3. Also contained in the appendices are a step-by-step script development outline and two business proposals written to encourage video program development.

Before we actually get to the material at hand, let me briefly change the subject for one final thought. If Hollywood is your eventual goal, remember, as you read these pages, that corporate and theatrical script-writing are closely related in many ways. You might think of them as two different sides of the same busy street. Getting your start on the corporate side is an excellent way to learn the traffic conditions, stoplights, dangerous corners, shortcuts, and wrong turns: knowledge that will be invaluable on the day you decide to take the crosswalk to the other side.

PART I

The Scriptwriter's World

Broadcast Versus Corporate

Thousands of television programs air every day across America: news shows, documentaries, made-for-television movies, specials, music videos, health programs, and many others. Most of us are aware that nearly all of these programs require a script. We are also aware that millions of dollars are paid out annually to the people who write those scripts.

What many people are not aware of is the massive amount of television programming that people *other* than the general public view every day. I mean, of course, privately produced and aired corporate programs, television shows made specifically for an audience of millions of American businesspeople. Just like their publicly broadcast counterparts, these programs include news shows, documentaries, dramas, music videos, and health programs. Also like the public broadcast domain, millions of dollars are paid out annually to the people who write the scripts for these programs.

If public broadcast and corporate scriptwriting have so many similarities, what, you might ask, are the differences? Actually, there are several. Because money is often the first topic to come up in conversation, that seems like a logical starting point.

The Prime-Time Market

By some standards, the prime-time scriptwriter's market is more lucrative than the corporate one. As of this writing, the Writer's Guild of America, West (the West Coast writers' union) lists a minimum payment for a prime-time, half-hour story outline and script of $12,724.

Although this looks attractive at first glance, remember that that hefty paycheck must be weighed against a few other factors: first, just getting a job writing that prime-time show is extremely difficult. It usually takes years of beating the streets in Los Angeles or New York, honing your craft, stacking up rejection slips, and pleading with agents to represent your work. Second, once the ice is broken in prime time, the amount of work many free-lance writers actually get is minimal—one or two projects a year, if they're lucky. Also remember that the rate just quoted is for *prime time*. Scriptwriting assignments for most other types of broadcast programming pay much less. Finally, keep in mind that actually seeing what you write end up on the screen the way you wrote it is highly unlikely in prime time. Many situation comedies and dramatic television programs are written or revised by groups of staff writers who have to come to terms with the fact that the majority of their work will be rewritten by producers, directors, actors, and other writers.

The Corporate Market

A 10-minute corporate script pays an average of about $1500. Compared to the prime-time script by page length or running time, this is roughly one-third. Some corporate projects pay less, but others pay much more. Payment usually depends on things like the writer's credentials, the deadline, the complexity of the project, and the budget available to the client or producer.

Although these figures may not seem quite up to prime time, in this case, too, there are a few important factors to consider: first and foremost, getting the corporate scriptwriting job is much easier. I don't mean to imply that just anyone can write corporate scripts, but people in this field usually secure projects without the help of an agent on the basis of a personal interview and writing samples, and getting an interview with a corporate producer is much easier than meeting a prime-time producer. On top of this, corporate projects are much more numerous and geographically dispersed, with potential clients in virtually every major city across the nation. Finally, chances of seeing what you write actually get produced as you wrote it are excellent.

More Than Money

Money, however, is only part of a writer's reward. Probably more important to most of us is the fact that, like prime time, corporate scriptwriting is a highly creative and exciting way to earn a living. Someone who has not written for the corporate market might dispute this, but those who do write corporate scripts are keenly aware of it. If corporate projects didn't require high levels of skill and creativity, most writers wouldn't continue to focus their energies on the corporate market once they've worked in it. However, once writers become aware they can earn good money, flex their creative muscles, and see their work produced intact, they quickly become corporate believers.

The Real Differences

In the final analysis, there are actually two nonmonetary, meaningful differences between public and corporate scriptwriting: the working environment and the research factor.

The Working Environment
To someone not familiar with the business world, the inner workings of corporations can appear intimidating. Large companies often seem to have a kind of underlying tension and robotic orderliness about them. In reality, the three words that probably best describe a typical corporate environment are *bureaucratic*, *political*, and *conservative*.

Corporations are bureaucratic because they are structured like pyramids. Descending, broadening layers of departments usually function according to very specific rules and regulations. This layered structure often dictates that business must be handled through the proper lines of the organization and according to strict corporate protocol.

It's very important to meet commitments, such as dates and times. Also, follow up frequently to keep the client informed of a project's status.

—Chris Van Buren
Corporate Video Client

Corporations are also political. Regardless of the volumes of regulations, policies, and practices in any company, who you know and how you operate make a big difference. This means that many corporate managers want to have the "right" contacts, appear to be movers and shakers, and continually try to position themselves in more and more powerful circles. Corporations are conservative because they like their employees to look professional, act professionally, and function as team players who are supportive of the policies and corporate culture of that company.

Although many people knock this bureaucracy and "stuffy" culture, we must remember that corporations are also the engines that drive the most productive and profitable economy in the world. They provide solid career paths with lucrative benefit packages for millions of Americans. They are also rich with challenges and rewards for those with a passion for getting things done through other people and making large business enterprises grow and flourish.

Just what does all this mean to you as a potential corporate scriptwriter? If you want to be successful, you must gain a healthy respect for the corporate world and learn to function in the business environment with ease and comfort. It also means that, contrary to what some people might think, the business world isn't at all a bad place to work.

The Research Factor

Corporate scripts, unlike many of those written for prime time, require that the writer learn something about the corporation's people, policies, objectives, or work. This is because most corporate programs are informational, instructional, motivational, or a combination of these. They are produced to help viewers function better as employees and, as a result, keep the company's bottom line in the black.

These elements of information, instruction, and motivation may at first turn off some writers. After all, who would want to write a script about something like, well, like the probing techniques used by sales people to learn about their customers? One West Coast writer did, and he decided it didn't have to be a boring program. A portion of the story outline follows:

PROBING: A STORY OUTLINE

We open on a close-up of a door marked "Interrogation." Off camera a conversation is going on between two detectives, Doris McIntire and Gordon "Gordi" Hansen. Gordi is saying, "Why all the intellectual stuff! Just give me five minutes with this pigeon and I'll have 'im singin' like a canary!"

"How do I make you <u>understand</u>?" Doris replies, "these are <u>not</u> the old days! There happen to be much more effective ways of getting people to give you information!"

As the conversation continues, camera pans and widens to reveal Doris and Gordi. We see they are standing at a coffeepot in a police office. Gordi is a large man in his fifties. His grey hair and hardened features tell us he's a veteran cop with considerably more brawn than brains. His sleeves are rolled up, and a shoulder holster and gun hang at his side.

When interviewing for a first scriptwriting job, it never hurts to come into the meeting armed with some initial research on the company you hope to write for. This can often be obtained from a phone call to the public affairs department or through a visit to the business reference section of the local library. Having done a little "homework" in advance can be very impressive.

—Dick Jones
Corporate Video Department
Head

Doris, on the other hand, is an attractive, intelligent-looking woman in her late twenties. She is neatly dressed and wearing no gun.

Although it's obvious Gordi isn't going to be convinced, Doris continues to explain. "It's called probing," she says, "and not only is it ideal in our line of work, it's also a good management tool. Bosses can use probing techniques for effective communications with their people. It's even great for every day person-to-person communications!"

"Yeah?" Gordi says, pulling a pair of brass knuckles from his pocket. "Well, I ain't no mumbo-jumbo expert, but I got my own person-to-person technique. And my way's only got one management tool . . . pain!"

Doris isn't giving up. She tells Gordi that may be the old way, but today's experts use six simple techniques to probe effectively. TELL WHY, MENTION BENEFITS, BE SPECIFIC, ASK OPEN-ENDED QUESTIONS, LISTEN ACTIVELY, and QUESTION CLUES.

"You got any clues about the guy you're dealin' with in there?" Gordi says pointing at the Interrogation Room. "This guy's a three-time graduate of the Big House! He ain't gonna hear a word you're sayin'!"

"Come on," Doris says, heading for the Interrogation Room. "We'll just see about that." As she and Gordi reach the door, Gordi flips a light switch, and Doris reaches for the knob.

We cut to a blinding light clicking on, filling the frame. A reverse angle reveals a close-up of Vincent "Vini" Legatta, his hands thrown up in front of his face. Vini is a small, greasy New York type, chewing bubble gum.

Doris and Gordi enter as silhouettes and take seats across the table from Vini. As Vini's eyes are getting used to the light, Doris leans forward revealing her attractive face. Vini is obviously pleased. "Well, well," he says, "beauty . . . (squinting, trying to make out Gordi) . . . and the beast."

At this Gordi blows up and leaps to his feet. "Why, you little weasel, I'll . . ." Doris stops him. "Gordi, please," she says, "just trust me. You know how hard I've worked to set it up with the Chief so I can prove myself. Please don't ruin it." Begrudgingly, Gordi backs off and takes a seat again.

Doris now turns to Vini. "Vini," she says, "you don't particularly like interrogations, do you?"

"Actually," Vini says with a cynical smile, "I'd rather be sailing."

"Well," Doris responds, "I want you to forget that word. This is not an interrogation."

Vini glances around and blows a bubble. "Could'a fooled me," he says.

"This is a probing session," Doris says. "You see, you have certain information I need, and probing for information is something I happen to be a specialist at."

Vini gets a chuckle out of this.

Doris ignores him and begins. She tells Vini that she would first like to explain how the probing session they're about to have came about. The Chief, she says, is a lot like Gordi—he believes in interrogations . . . very physical interrogations. "But after months of persuasion, I've managed to convince him to give me a try at a more effective and humane way of getting information—probing." She's determined, she says, to show him it works.

Vini likes this. He says, "You mean you can promise me Bigfoot over there won't get involved?"

"Exactly," Doris says.

Vini grins. "Okay . . . Okay," he says. "And just what is this, ah . . . probing . . . stuff all about?"

Doris tells him it involves six simple steps that make for effective two-way communications. She continues to explain. (Titles are supered as she discusses each step.) "The first step," she says, "is to TELL WHY I need the information."

"How come?" Vini asks.

Doris explains that telling a person why you want information does several things. First it removes any suspicions or mistrust the other person may have. This, in turn, gets rid of any barriers up front, builds trust, and opens the way for clear, mutually supportive two-way communications. "All in all," she says, "it starts the session out on a positive, cooperative note."

"So in this case," Vini says, "you'd tell me something like there's been this crime committed, and you'd like any information I could give you to help solve it. Right?"

"Right," says Doris.

"Yeah," Vini says. "I like that. But then, I might ask, What's in it for me?"

"That comes next," Doris says, "MENTION BENEFITS. This lets the person giving the information know that there is something in it for him." She explains it could be an actual reward of some kind, or just the fact that they are being a great help to the person probing. The point is, it makes the person want to give you information.

"You mean I get benefits from bein' a fink?" Vini asks.

"You certainly do," Doris responds. She explains one very definite benefit is the fact that if he is guilty, she'll recommend that the judge go light on him. She also mentions that although he may not have experienced the feeling in a while, just getting things off one's chest can work wonders for the conscience. "Probably the most direct benefit in your case, though," she says, "is the fact that Gordi here won't be allowed to rip you to pieces."

I'd call that a pretty creative way to handle a "boring" subject. Granted, many corporate scripts are less elaborate. Do corporate scriptwriters mind this? Rarely. In fact, most writers seem to find that just the process of learning some supposedly boring subject often turns out to be extremely interesting. Then, finding a creative way to communicate what they've learned to the employees in a company becomes an even more challenging and rewarding process.

Career Paths

Another question you may have about corporate scriptwriting is exactly where such a job eventually leads. There are a number of possibilities.

Staff Writers

If you land a job as a staff scriptwriter in a large company, you will probably earn approximately $40,000 per year and a solid benefits package. After a few years of good service, you'll be able to progress in one of two ways.

Every good writer is curious — curious about people, emotions, products, processes, statistics. Curiosity makes it interesting to write about subjects from substance abuse to earthquake preparedness, from fiber optic technology to customer relations skills, from new product introductions to back injury prevention. Curiosity is one of the writer's most valuable tools.

—Patti Ryan
Corporate Producer

If you move up in your own department, chances are you will eventually supervise writers or direct and produce corporate programs as well as write them. Having taken this path myself, I can say that although I wouldn't have thought so 10 years ago (I felt writing was the only creative activity anyone should ever consider worthwhile) directing and producing are also extremely creative and rewarding, and now I enjoy them nearly as much as I do writing. Still farther down this same road, if you turn out to be management material, you may end up running the corporate video department and loving it.

If you move up the ladder and out of your department, areas of your company like public affairs and employee communications are often close corporate relatives of the video production group and natural avenues for the same type of career progression just mentioned.

These departments usually produce such publications as company newspapers, brochures, press releases, executive speeches, company bulletins, public service messages and others. They also represent the company to the news media, to other companies and various agencies in the community.

Free-lance Writers

Many companies hire out their scriptwriting work to free-lancers. If you take this path, you will write for different companies on an independent, per-project basis. Chances are you will eventually move into producing and directing as a part of the services you provide or move into other types of writing such as screenplays and teleplays for the entertainment industry or perhaps books or print work for general publication.

Whether your future lies in a corporate office or on a Hollywood soundstage, your tenure in corporate scriptwriting will be a big help in getting you there.

Finally

So what's the final word? Scriptwriting for the corporate market is creative, potentially lucrative, and personally rewarding and a challenging, exciting career path for those who are willing to give it their all.

What will this book do for you toward achieving success in this field? Obviously a book can't provide someone's "all." The personal drive, determination, and plain old hard work are up to you. What this book can provide, however, are two things that will perhaps make the process a little quicker and easier: a solid foundation on which to start building your career skills and, I hope, a shot or two of adrenaline to get you started.

The Corporate Script

2

□ □ □ □ □

Blueprint for Production

Before you are ready to write corporate scripts, you'll need some definitions and a sense of the scriptwriter's perspective, that is, a feel for where you and your script fit into the production process and how you affect the people involved in that process. To help you begin to gain that perspective, let's start with the most important definition of all: What exactly is a corporate script?

The Corporate Script: A Definition

The words *blueprint*, *framework*, and *skeleton* are often used to describe the essence of a script. Although all three are accurate, for the purposes of this book, here is another, two-part definition of a corporate script:

1. a precise, written description of an imagined series of events and audiovisual elements, expressed in terms of standard industry terminology and formats
2. a product

A precise written description refers to the clarity and visual quality your script must contain. For the client and producer to be able to visualize and subsequently approve it, they must be able to imagine what you've written taking place on the screen.

Series of events and audiovisual elements refers to the individual scenes and parts of scenes that you, as the writer, put down on paper. They are *events*, such as a host walking into a room to address the camera or two characters sitting in a coffee shop discussing the company's latest safety procedures. They are *elements* like dialogue, stock footage, graphics, titles, dissolves, music and sound effects, in other words, all the parts that make up the script as a whole.

Expressed in terms of standard industry terminology and formats takes into account the mechanics of how the script is written. To be recorded, these events and elements must be put on paper in a way that the production and postproduction teams can use.

The second part of the definition, *a product* reminds us that a script that may be a literary creation is still a product for sale. This means that the producer, who is paying for it, has the ultimate right to request changes to it as she sees fit. Keep in mind that these changes may not always be—and in fact many times are not—what the writer would prefer to do. The successful corporate writer learns to accept this fact. He also makes a point of never getting so "close" to his material that he cannot objectively view or change it at any time.

Film and Videotape Production

A script is one of the first steps in a complicated and expensive production process. That process is made up of four parts:

Writing   research and script development
Preproduction   preparations needed to accomplish production and postproduction
Production   the recording of the script on film, videotape, audiotape, or a combination of these
Postproduction   editing the recorded pictures and sound into a completed program

In order to gain a solid perspective of how your script fits into this process, let's look more closely at each phase.

Writing

This is your part. It usually begins when the producer gets a request for a program from a client. In most cases the client has some communication problem and a film or videotape appears to be the most effective solution. In many cases, before calling in the writer, the producer will meet with the client, develop a program needs analysis, and verify that a program is the right communication tool for the job.

If the need is legitimate, you will then get the call to take on the project. Your first order of business will probably be a meeting with the producer to discuss the project parameters and fees.

In many cases, you will be told that payment will be on a milestone basis, which means that you will be paid installments as the various drafts are completed and turned in. A common milestone schedule would be one-third of the payment as an advance, the second third when the first draft of the script is delivered, and the final third upon delivery of a client-approved shooting script. Getting to this final payment often takes about three drafts.

Following the initial producer meeting (maybe on the same day), you will attend another meeting. This one will probably include you, the producer, the client, and perhaps some content experts. From meetings like this and other interviews and field visits, you will most likely develop two documents: a treatment and a script. In some cases, the producer may also want you to develop a content outline and maybe even a program needs analysis or something like it.

How many documents you are required to write usually depends on the project. Technical, instructional, and complicated projects often require a content outline to assure the client that the facts have been properly researched and structured. Motivational programs often have very little hard content and thus may not require an outline. Another factor is simply the producer's way of developing a script. Some want needs analyses, outlines, and treatments; others want a script and nothing more.

Figure 2.1 A client meeting in session. Writer is in foreground, closest to camera.

Following research, you will enter a conceptual phase in which you brainstorm concepts, story lines, and other visual methods for communicating the message as effectively as possible. This phase will result in a treatment, which is a simple narrative of the program. Finally, after the treatment is approved, you will write the first, second, and third drafts of the script.

All this work will take about 5 weeks and allow both the client and producer to monitor, visualize, revise, and finally approve the script you've written. It will also allow the producer to create a budget, which will accurately predict the cost of the entire project.

At some point, take a moment to review Appendix 1 for a more detailed, step-by-step explanation of the entire scriptwriting process.

Preproduction

Once the script is approved and budgeted, your presence may seldom be required, if at all. If the producer isn't directing the project herself, a freelance director will be brought in and probably an assistant director or a production assistant. The director may have a question or two for you about how you visualized a certain scene, or the client or producer may want a quick, last-minute revision.

With or without your presence, your script will become the focus of a great deal of attention. Preproduction is a time for making arrangements and attending to all the details necessary to bring off an efficient, well-organized shoot.

The entire production must be set up and precisely scheduled. The amount of time involved in this setup process is directly related to the complexity of your script. If it has seven speaking roles, seven actors will have to be cast, which means looking through pictures and casting agency books, calling agents on the phone, scheduling auditions, or perhaps arranging for employees to play some of the parts.

If your script is played out in five remote locations, each place will have to be located, scouted, and analyzed to determine whether it will meet the needs of the production team in terms of sound, light, equipment, and personnel considerations. If it meets all these criteria, its use must be arranged and scheduled. Getting special insurance riders, city permits, and perhaps fire or police officers may also be required.

If your script calls for an elaborate set, it will have to be designed, built, and lit. If it calls for a great deal of artwork, an artist will have to be brought in to create the drawings, paintings, or electronic art.

As you can see, then, what you put down on the page is not to be taken lightly. Many hours and perhaps thousands of dollars will be spent making sure all the elements you've included can be brought together and recorded accurately.

Figure 2.2 A studio camera operator lines up his shot in preparation for an actor's entrance.

Production

When all the arrangements are made, you can be assured they will have been scheduled into the most economical order possible because the production process is complicated and expensive. Your script will be shot totally out of order, and often elaborate lighting and sound setups will be required to record each scene convincingly.

A crew of perhaps three to seven people will be hired to perform specific tasks during production. The crew might include:

Director　the person in charge of all aspects of recording the script on film, videotape, or audiotape

Assistant director (AD)　the director's right-hand person

Lighting director/camera operator (LD)　the person in charge of the main camera work and lighting on the program

Sound engineer　the person in charge of recording sound on the program

Gaffer　the lighting director's first assistant

Grip　a general labor assistant

Production assistant (PA)　a general runner and "details" assistant

Each of these people will put in a great deal of time and effort and sometimes endure great pressure and frustration to make sure the scenes you have written are recorded properly. To illustrate this point, let's briefly examine a typical scene a corporate script might contain and see what could be involved in getting it on tape.

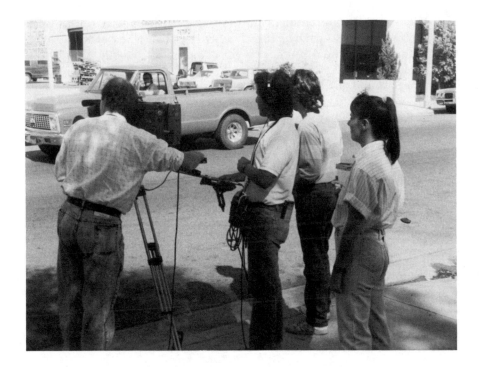

Figure 2.3 A small corporate crew on location in an industrial area. A truck entering a parking lot at left is being recorded. Program was on parking safety.

EXT PARKING LOT—SUNSET
Bill comes out the side door of the building and gets into his car.
He starts it up and is about to drive off, when Leslie pulls up
beside him. She rolls down her window and says . . .

                    LESLIE
          Headed for dinner?

                    BILL
          As a matter of fact, yes. We're all going
          over to Alphonso's. Wanna come?

                    LESLIE
          Sure. I'll meet you there.

Both Leslie and Bill now back out onto Fifth Street and drive out
of sight around the corner.
                                        DISSOLVE

Quick and easy? Not really. This "brief" scene contains a number of potentially frustrating and costly complications.

The first lies in that initial scene heading: "SUNSET." Shooting videotape or film at this time of day is almost always a frantic situation because both the volume and the color of daylight are changing by the minute. For videotape cameras this means constant white balancing (referencing the camera to the color white) and adjusting the intensity and color temperature of additional lights being used. To accomplish this, someone must run in front of the camera with a white card, probably after each take, and someone else must continually make adjustments to each light.

The second major problem in this scene involves sound. If as written, it takes place in a parking lot, there will probably be traffic close by. Traffic means continual retakes to get actors' lines "in the clear" from background noise.

These sound and light problems are multiplied because the director must shoot this scene from several angles over a period of hours, and these angles will all have to appear to have happened at the same instant.

Timing will also add to the problems here. Getting Leslie to pull into the camera's frame just as Bill is about to pull out will take a few tries. Then there's the problem of backing onto Fifth Street and pulling away. If this action has to happen simultaneously with any of the lines, timing again becomes a problem. An open spot in traffic must occur at the moment when Leslie says her line and the two back up.

I bring up these issues not to discourage you from including the proper action in your script and having it take place at the proper time of day in the appropriate location. You should be keenly aware, however, that when you write simple words like *sunset* or *parking lot* into a scene, you could be creating hours of work and expense, when another time of day or place might have done just as well.

You can take several steps to avoid these types of problems. First, remember that the more you can pick up about the production process in general—and, better yet, the methods of the producers you write for—the more you are increasing your chances of writing good, producible scripts. Toward this end, if the producer will allow it, you should jump at the chance to ride along on shoots to see the process firsthand. If the producer doesn't seem crazy about the idea, offer to act as a free production assistant. This will probably do the trick.

Another helpful practice is to take some time to look over every scene you write and remind yourself that if it's approved as is, everything on the page will have to be created or located, brought to a location, and made to happen properly timed in front of the camera, several times. Then ask yourself if the same point and impact could be achieved with a less elaborate or complicated scene. If the answer is no, the scene should remain as written, complicated or not. If the answer is yes, the scene should probably be rewritten.

In our example, the key point is whether the conversation has to take place at sunset in cars in a parking lot. Could Bill have just as well been wrapping up things at his desk or getting his coat off the rack when Leslie walked up and the same conversation took place? Would this suggestion that it was near the end of the workday have sufficed instead of an actual sunset? If so, the result would have been a scene shot in 1 hour instead of 3.

Postproduction

With all the footage in the can, the program is ready for editing, which is the process of putting together all the out-of-order bits and pieces that have been recorded and making them into a cohesive program. To help

Figure 2.4 An on-line editor at work on a corporate program. Production script, with notes from the shoot, is on the console in front of him.

facilitate this process, the editor will probably use a copy of your script on which an assistant director or production assistant has made notes. This master script lets the editor know which takes were good and which were bad.

After the program is edited, duplicated, and distributed, a master file may be created. The script will remain in it for consultation in any future revisions or other references to the project.

The Skills Required

To research and write good scripts that also help to facilitate the production process you've just been through, you'll need the following business and writing skills:

Producer/client handling skills These will make you effective at working with corporate employees in business situations.

Research skills These skills will hone your ability to dig out the proper facts from the resources available.

Concept thinking skills Concept thinking skills will help you open up your creative channels and allow good ideas to emerge from the information you've uncovered.

Visualization skills Visualizing helps you imagine very clearly what every scene you write will look like on the screen.

Format skills These skills will teach you the basic structure and terminology present in every script.

Dialogue and narration skills These skills will teach you to create words and speech patterns that flow naturally from actors' mouths and at the same time tell your story with maximum clarity and punch.

Structure and transition skills These skills will enable you to arrange and present information in a way that makes it easy to follow and thus quickly absorbed.

In the following chapters, we will explore each of these basic skills in much greater detail.

PART II

The Scriptwriter's Basics

Producers and Clients

VIPs

The producer and the client are the two most important people in a corporate scriptwriter's life. It's critical that you understand how each one thinks and works and how your creative services help accomplish their corporate goals.

Producers are usually on staff in the company that hires the writer. They are typically middle-management employees whose primary job is to make successful film or video programs for their clients. *Successful* means programs that solve the client's communication problem within the time, monetary, and political constraints of the company. Toward this end, producers usually supervise a project from the day the idea is proposed until delivery of the finished program. Many producers have years of experience and prior background in television production. When a project becomes active, one of the first and most important things a producer does is hire a writer.

Clients are also on staff. They, too, are typically middle-management employees, but they are the ones with a communication problem to be solved. In a sense, a client "hires" the producer to solve this communication problem. Although both may work for the same company, *hire* is an accurate word because clients often pay for their production out of their departmental budgets. One of the first and most important charges a client incurs is the writer's fee.

As the writer, you will work for both the producer and the client. You are hired and paid by the producer, but you will also take instruction from the client. Your main objective is to write scripts that meet with the producer's approval and at the same time satisfy the needs of the client.

With this brief overview of the producer-client-writer relationship in mind, let's now get a much closer look at producers and clients.

Producers

You will find that the personalities and operating styles of corporate producers vary widely. Some are firm and decisive, but others vacillate continually. Most are truthful and direct, but there's always the one who will skirt the issues. All this means that, just like the rest of us, producers are *human*.

Dual Perspective

Good corporate producers keep a dual perspective on every project they manage: creative and business.

Gain your client's trust as a professional businessperson, and work hard to keep it. This is one of your best investments in your future as a scriptwriter.

—Dick Jones
Corporate Video
Department Head

On one hand, they would often like to give you total creative freedom to try whatever wild ideas you might come up with and take as much time as you need in developing them. This is their *creative* perspective.

On the other hand, they are usually under considerable pressures from their company. Most times, these pressures do not allow them — or you — to enjoy the luxuries just mentioned. These pressures are things like budgetary constraints, time squeezes, and political considerations. This is their *business* perspective.

Successful producers learn to balance these perspectives carefully. That balance gives their creative people enough freedom to function well,

SCRIPT WRITER EVALUATION				
NAME			Date ____	
PREVIOUS CLIENTS / PRODUCTION COMPANIES				
	Weak	Avg	Good	Ex
INNOVATIVE / CREATIVE APPROACH VS. PREDICTABLE	—	—	—	—
DEPTH / THOROUGHNESS / GRASP OF MATERIAL	—	—	—	—
GOOD RESEARCHER	—	—	—	—
ASKS RIGHT QUESTIONS	—	—	—	—
UNDERSTANDS OBJECTIVES	—	—	—	—
VISUAL VS. LITERAL	—	—	—	—
WRITES GOOD DIALOGUE / DRAMATIC SITUATIONS	—	—	—	—
ECONOMY OF WORDS	—	—	—	—
WRITES TO AUDIENCE LEVEL	—	—	—	—
ENTHUSIASM / COMMITMENT	—	—	—	—
DEADLINES / SPEED / HARD WORK	—	—	—	—
WILLINGNESS TO MAKE CHANGES	—	—	—	—
WILLINGNESS TO FIGHT FOR QUALITY	—	—	—	—
HANDLES PRESSURE	—	—	—	—
ABILITY TO WORK WITH CLIENT / PRODUCER	—	—	—	—
AWARE OF COSTS OF PRODUCTION	—	—	—	—
KNOWS MEDIUM - VIDEO / MI	—	—	—	—
KNOWS MKT / SALES / TRAINING	—	—	—	—
SUBJECT MATTER BACKGROUND	—	—	—	—
INTEGRITY / CHARACTER	—	—	—	—
SENSE OF HUMOR	—	—	—	—
OVERALL EVALUATION	—	—	—	—

Figure 3.1 One producer's scriptwriter evaluation form.

while allowing them to keep enough control over the project to bring it in on budget, on time, and to the client's satisfaction.

It is in your best interest as a writer to be aware that this balancing act exists and to make it as easy as possible for the producers you work for. A writer who begins to make a producer's life difficult by being obstinate about changes or by making unreasonable creative, monetary, or time demands immediately becomes less likely to be called back on the next project.

Loose-End Haters

In addition to the balancing act, most corporate producers juggle multiple projects at one time. Remember, too, that each of those *multiple projects* is made up of *multiple facets* in each of the *multiple phases* it must pass through on its way to completion. If all of a producer's "multiples" fell right into place on every project, the job would be a cakewalk. Unfortunately, that rarely happens. What usually happens instead is that during this juggling act loose ends keep developing.

As you might guess, these loose ends are a constant bother. If producers have to spend a lot of time picking up the loose ends of certain writers, you can bet they will remember those loose ends when they pick up the phone to hire someone.

Respect

Most producers have a difficult time dealing with writers who either aren't aware of the producer's dual perspective or who are too inflexible to adapt to it. Writers who respect the producer's position stand a much better chance of staying high on the list, even if they disagree once in a while or ask for some concessions.

The important thing to remember is that respect does not mean fear or compromising values. You needn't be a yes person, just a perceptive and flexible one. If you run into a situation where a producer insists that you go beyond these two traits, I would suggest looking elsewhere for scriptwriting work.

Memories of Successes

Finally, good producers remember their successes and the people who helped achieve them. If you are a writer who contributes to success consistently, you will surely have trouble handling the volume of projects you're bound to be assigned.

Clients

Writers and clients have stereotypical images. The writer's image is what could be called Hollywood spacey. They are disheveled and absent-minded. They are poetic and visionary. They are frequently unshaven and always free-spirited, with a tendency to show up late for meetings or to call at the last minute and say they need to meditate before they can deal with reality.

The client's image is very structured and stuffy. Clients are immaculate clones in three-piece suits. They carry leather briefcases, are

The corporate scriptwriter should always dress appropriately for the corporate culture he or she is working in. This may mean a suit and a tie in some situations and blue jeans and a sweatshirt in others.

—Dick Jones
Corporate Video
Department Head

totally helpless without rules, regulations, and pocket calculators, and leave your meeting to go patronize some arrogant vice president.

One of the most striking points about these two images is that they are totally *dis*similar. According to stereotypes, writers are absolutely nothing like corporate clients, and both should regard each other with a certain amount of suspicion and contempt.

Because these stereotypes can persist, especially with clients working on their first programs, one of your first jobs as a writer is to dispel this Hollywood spacey image.

In order to do your job effectively, you must create a trusting, open, and, above all, *comfortable* relationship between you and the client from the very first handshake. If you accomplish this, the client will be willing to place the responsibility for the project in your hands. If you appear to be totally out of place in the corporate setting, your client will be worrisome, tight-lipped, and very possibly suspicious of the producer who brought you in.

Making the Impression

You can dispel the Hollywood spacey image by paying careful attention to two things: dress and manner.

The proper dress Always wear a business suit or the *appropriate* dress and nylons or slacks, tie, and sport coat for a corporate meeting, not the "Miami Vice" nightclub look or the newest fashion statement from Paris. Also be sure to come neat, combed, and clean-shaven (or with a neatly trimmed beard or mustache). Make sure your shoes are polished, your nylons are free of runs, and always carry a briefcase. Inside the briefcase, always keep

- an assortment of pens and pencils
- a professional (preferably leather) notebook full of paper
- a portable tape recorder with several cassettes and spare batteries
- a pocket calculator
- your business cards

Figure 3.2 Business card used by a California writer/producer. Letterhead paper, custom envelopes, and special printed inserts are also a part of this person's self-promotion package.

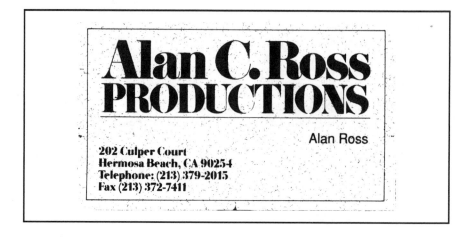

Client Profile: The Attentive Type   Case Study 1

There were five of us in the room: the security manager, two of his special agents, myself, and the producer. The gathering was an initial meeting for a program on security measures being implemented to curb substance abuse in a large food company.

Because I was going to be writing the program needs analysis, the producer and I wanted to leave the meeting with some good, solid design information: audience, objectives, utilization, and so on.

After I had explained this to the security manager, I asked him my first question: "Tell me about the audience. How much do they know today about security measures being used against drug abuse?"

He thought for a second and then said, "Not much, I don't think. At least the guys up in the northern division didn't know much when we caught them red-handed drinking beer at lunch." He then turned to one of his agents and continued, "Remember that one, Dell?"

"Boy, do I," Dell said. "Back in 88, wasn't it?"

"Yeah. That one guy had just reached into a cooler. Then there was the bust we did on the three guys from Phoenix."

The other agent chimed in, "They never knew what hit them."

This conversation began a series of recollections that went on for roughly 15 minutes but gave me almost nothing I needed to know. I tried to be tactful, saying, "Those are some interesting stories. We'll be talking more about those kinds of things when we get into the content research on the project. Right now, though, I'd really like to focus on things like the objectives and the audience. I realize they're not the most interesting aspects of the subject, but they really are important. Now about that audience knowledge?"

"Yeah," the director said, "Objectives are important. You know our objective is to have a real hard-hitting video, Ray. I remember one bust that would make a perfect story, by the way. It started on a Friday evening when we got a call from a hysterical employee. Remember that call, Arnie? Down in San Diego?"

"Sure do," Arnie said. "Boy, that was a humdinger!"

And so it went for another 15 minutes. Two hours later, the exasperated producer and I left the manager's office with enough stories to write a prime-time series, but virtually no design information whatsoever.

We went to a local coffee shop and plotted strategy. After a good deal of discussion, we finally decided that the way to get what we needed from this particular individual was to get him *alone*, so he would have no one to reminisce with.

We set up another meeting a few days later, and sure enough it worked. Without "the boys" handy to conjure up one story after another, the director got down to business and actually gave us what we needed.

The lesson? Client handling can be a big part of doing effective research.

*M*ake the client
feel you are a fellow
employee after the
same company goal.
The writer should not
be viewed as an
outsider.

— Chris Van Buren
Corporate Video Client

The corporate manner   In any corporate meeting, you should always present a polite, inquisitive, and confident image but *never* display an egotistical or laissez-faire attitude.

Remember, the project is important and involves thousands of dollars. Chances are the client's boss has delegated the responsibility to the client for completing it promptly and effectively. The client's professional reputation may be on the line. You must project yourself as a person who is both competent and enthusiastic about taking on the responsibility for not just preserving but also *enhancing* that reputation through the use of your creative skills.

This type of dress and manner, incidentally, should continue in all business interactions throughout your corporate writing career. The exception would be if you find yourself in a business situation or environment where typical corporate standards are out of place. If there is ever a question in your mind, go with the conservative, corporate image until you find out otherwise.

Client Profiles

Another of your important objectives is to be able to spot quickly the type of client you may be dealing with. This allows you to adjust your manner and presentation accordingly. Among the wide variety of personality types you'll encounter in the corporate world are several predominant ones:

- undiscovered writers
- busybodies
- instant decision makers
- yessirees
- committee heads
- vacillators
- ramblers
- plain old perfect clients

Let's take a closer look at each of these client types and some of the skills you'll need to work effectively with them.

Undiscovered Writers

These clients have never in their lives written a script, but they know your business better than you do. They may be in a "cousin" field to corporate program production such as public affairs, employee communications, or training.

Undiscovered writers assume that scriptwriting is the same as writing a newspaper story, a company bulletin, or a training course script for an instructor. If you try to explain to them that print writing is to be read, course scripts are to be delivered live, and program scriptwriting is something totally different, your message often doesn't register. Instead, they look up from their notebooks in the middle of content meetings and make comments like

- "We need to do a montage with music in this area."
- "A dramatic vignette is what we'll put here."

- "I think we'll build this around a series of interview segments."

Your best bet in working with these people is to acknowledge their comments tactfully as great *suggestions* that warrant serious consideration at a later time. Then, later, if they really are good suggestions (sometimes they actually are), by all means use them. If not, chances are by the time you meet again, they'll have forgotten what they originally wanted and love what you've come up with instead.

When dealing with undiscovered writers, you must *never* come off as too aloof to consider their suggestions. At the same time, however, you should not be locked into any one concept at a research meeting. This is a time for flushing out the information needed to eventually make those concept decisions. When the time comes, you and your producer are the ones who should be making them.

Busybodies

These overworked managers want a script pulled out of a hat. They have half an hour to sit with you in an initial meeting, but after that they'll be in Chicago next week, up to their eyeballs in reports the following week, doing sales seminars on the road for the next two weeks, in training, revising the budget, and writing performance reviews. As luck would have it, they happen to be able to squeeze in 30 more minutes with you in 2 months—to approve the final program!

What these people don't realize is that writing scripts takes more than a scriptwriter and a producer. It takes client input, support, and approvals at many places along the way. It also requires that content experts provide information or at least the proper guidance and contacts to get that information. It frequently requires someone to open doors to other departments and to act as a liaison to possible interviewees, executives, and other sources.

You should, again, *tactfully* let your busybody know that. Make the point that the scriptwriting process is really a collaboration. Your job is to supply the organization, conceptualizing, and writing skills. The client or a person the client delegates must provide the content and approvals *throughout the process* and not just on the final approval day. Also, be sure to solicit your producer's approval in acquiring that support from clients, content experts, and anyone else involved with script development.

Instant Decision Makers

These employees want to *appear* to be in complete control at all times. When they're given a choice about a script element, they'll often make an instant decision, with no thought whatsoever as to whether it will really work.

As an example, I was once describing to an instant decision maker how a series of interviews might work into a script concept I was considering. I said, "We could do them as scripted pieces with each interviewee looking directly into the camera, or. . . ."

*T*ry to learn as much as possible about your client— his or her goals, motivation, interests, etc.

—Chris Van Buren
Corporate Video Client

He stopped me there. "Yes," he exclaimed, leaning forward in his chair, "that's *exactly* what I want. Looking directly into the camera. Eye contact. Perfect!"

As it turned out, the interviews were eventually written and shot in a totally *un*scripted fashion with the interviewees looking slightly *away* from the camera at an interviewer. Had they been done the way I first imagined, they wouldn't have worked nearly as well.

Your tack with instant decision makers, just as with the undiscovered writers, should be to pacify them for the moment with a tactful comment about how great their ideas sound, but take those ideas with a grain of salt. Consider them later, when you've had a chance to absorb all the design and content material and then decide what works best.

Yessirees

These clients seem too good to be true. Everything you say sounds perfect to them. Tell them you're considering a role-play scenario and they'll tell you what a wonderful idea that is. Five minutes later, say you're considering a documentary approach, and they'll follow right along, saying that does actually seem like a much better way to go. Later pitch a music video, and watch them start picking song titles.

The fact is these clients have *no idea* what they really want. It's up to you and the producer to lead them down the right path. The important thing to remember with the yessirees is not to base your script decisions on what they think they want. Instead, base your decisions primarily on design and content research. That way you can't go wrong.

Committee Heads

The people who speak for committees are often the opposite of instant decision makers. Getting them to make a decision can be like pulling teeth because they represent a group of people who all have a vested interest in the project and who all want a say in the decision-making process. As a result, committee heads want to bring five, seven, or maybe 15 people to your meetings and have a roundtable discussion about every word in the script.

This is sometimes unavoidable, but the problem is that it takes forever. One person feels humor is right, and the other despises laughs in an instructional videotape. Another wants every detail on the subject included in the script, and still another feels only an overview is needed. One wants actors, and another wants employees to play the roles.

Try to avoid script approval by committee by stressing to your committee head that although various people may need input into the script, following two golden rules will help the project along much quicker and lead to a better end product:

1. *The client or a person she delegates should be the single point of contact for the writer and producer.* This means all input from others with a stake in the project should be filtered through this contact person. In this way, a good deal of redundancy and unacceptable changes can be weeded out *before* they ever get to your script meeting.

Scriptwriting by Committee   Case Study 2

I remember sitting in a large business conference room at one end of a horseshoe table arrangement. The other 15 or so seats were taken up by clients, content experts, administrators, instructors, and the producers of a series of four public service scripts I was writing. The subject was nutrition. They each had a copy of my first drafts in front of them, and this was the time for sharing their input.

I had found out earlier that day that the producer and client were very happy with the scripts, but their hands were basically tied by the desires of this committee. To make matters worse, there were power struggles going on daily.

As the meeting began, I sat poised with my pen and notebook ready. Over the few hours, I heard conversations like the following.

"I love the section on the four good groups, but shouldn't exercise precede it?"

"Absolutely not, Marion. The overall message of basic nutrition is the primary objective here: basic food groups, healthy eating habits."

"But Daniel, exercise is a part of basic nutrition. I think we all agree on that, don't we?"

"I don't know about the exercise, but I'd love to see an opening with shots from a market, lots of fruits and vegetables. You know, a real upbeat thing on food, and what better place than a grocery store?"

"You know, we could have kids in it too. I mean, kids are part of this audience."

"My daughter's available!"

"Personally, I like the idea of animation. Is is too late to write in animated sections, maybe to just replace certain parts of the scripts?"

"I think we're getting away from the point here. Let's get back to the fruits and vegetables."

"Is your preference animation or live action, Phil?"

"Is someone taking notes?"

It's called approval by committee, and it's the worst nightmare for any writer. Everyone gets into the act, everyone is an "expert" on scriptwriting and motion picture photography, and all the work you've done simply gets politically maneuvered into oblivion.

Because companies are bureaucratic and political by nature, approval by committee can't always be avoided. As a writer, however, you can certainly try to encourage a one-person contact arrangement. One key person meets with whomever else is necessary, hashes out whatever changes are proposed, and presents them to the producer in a one-on-one meeting. Hopefully, the producer has final say to be sure the changes are really in the best interest of the project. When everything is decided, you are called in to revise as necessary.

2. *The writer and producer should be given the authority to make the final decisions regarding what changes get included.* You and the producer are the media experts. As such, you should be empowered to include and exclude whatever input you feel is appropriate for the good of the project. If your client trusts you both, this should be no problem. If he doesn't trust you, you need to somehow gain his trust as quickly as possible.

Vacillators

These clients are probably worried about their ability to actually deliver an acceptable final product. Because of their lack of self-confidence, rather than make a decision on some part of the script or perhaps even the *entire* script, they may ask that you do it "both ways." Their excuse will often be that they'd prefer to compare ideas to be sure they get the best one.

These clients will take whatever you've written back to their offices to "review it carefully." That careful review usually means running it by the boss to test the water. If the boss likes it, the vacillator will also like it. If the boss comes out of his office wearing a scowl, the vacillator will be back to you saying that *he's* decided he wants a new version pronto and maybe two, just to be safe!

Your best (and frankly limited) strategy against vacillators is simply to be totally professional and keep reassuring them everything will be fine. Then, write them an excellent script. With one positive experience under their belts, they may be inclined to exercise a little more healthy risk taking on future projects.

Ramblers

No, these are not old cars; they're clients who cannot seem to focus on the topic you're trying to explore. If you tell them you want to discuss objectives, their first few words may relate to that. Then, something will cue a recollection or some other subject, and they're off rambling for 10 minutes of nonstop gobbledygook.

Like vacillators, ramblers can be difficult to deal with. Most times your best bet is to simply sit back and wait them out. When you see an opening along the way, try to steer the conversation back on track. Eventually you'll get what you need if you're just patient enough.

Plain Old Perfect Clients

The kind of people you will probably work with more than any other are plain old perfect clients. They are enthusiastic about the project, and they are more than willing to dig in and provide you with whatever you request. They have content experts at your disposal, audience members to interview, job locations to visit, telephone numbers, names, good ideas, and as much time and energy as you need from them. They're capable of making good, solid decisions when you need them, but they also have a keen sense of when more scrutiny or a higher approval is needed.

Be alert for hidden agendas and try to understand them. Programs aren't always being done for the reasons stated. Maybe the department manager needs to prove to his boss that his idea is better than a competing department's.

—Alan C. Ross
Corporate Writer/Director

Perhaps the best thing about plain old perfect clients is that they regard you and your producer as the experts in your field. As such, they trust your decisions and in most cases recognize when they are starting to overstep their boundaries as clients.

The result is that you get plenty of creative fuel in the form of support, approvals, and input, but you and your producer always remain in the driver's seat, which is a nice feeling. The only real problem with plain old perfect clients is fighting off the urge to adopt them as permanent clients on every script you write!

A Few Final Thoughts

Remember that there are often *degrees* of the client and producer traits we've been discussing. A client may be great to work with in many ways, but she may make too many on-the-spot decisions. She may want her boss to approve everything you do, even though she makes good decisions herself. A producer may allow you adequate creative freedom in some areas but be too restrictive in others. Working with these people and sensing the degrees of tact and strategic maneuvering necessary in each situation is just another of the many challenges that make the corporate scriptwriter's job so rewarding.

I hope you now have a reasonable sense of something I mentioned earlier: scriptwriter's perspective. You should know what corporate scriptwriting involves and exactly what a script is. You should also understand how the script fits into the production process, and you have just been introduced to the key players in your part of that process: clients and producers.

With this information under your belt, our next step is to get down to the business of writing.

A good writer can take my needs as a corporate executive and the producer's needs as a video creator and mix these with his own skills and creativity to come up with an effective script.

—Tom Anderson
Public Affairs Director
Corporate Video Client

■ ■ ■ ■ ■ ■
You Script It 1

During your initial meeting with the producer and client, the subject of approvals comes up. Your client says he's going to be very busy for the next month with a road trip. He goes on to say he really has no concerns about whatever you come up with. He tells you and the producer just to move ahead with the project and when he gets back he'll look at the script.

As far as the approval process is concerned, he says he would like to run the final draft of the script past his boss and the committee he chairs. This, he says, is to get their input in case there are any minor, last-minute revisions. He says he would like you to attend that meeting to get their reactions and take notes.

What should you say or do?

4

□ □ □ □ □

Program Design Research

The Importance of Research

Research is the process of gathering, exploring, understanding, and organizing all the information involved in your scriptwriting project.

Writing the script then becomes simple, creative, and fun. Without this understanding and organization, however, scriptwriting can be frustrating, complicated, and time-consuming. Writers who do not do proper research end up spending more time and energy trying to figure out what they need to communicate, and how and why, rather than on the enjoyable creative processes.

To avoid this, it's important that during your research you learn to gather two distinctly different, but equally important, types of information: program design information and content information. We will look more closely at content information and how it is researched in the next chapter. For the moment, let's focus on design.

Case Study 3 — *A Design Allegory*

Imagine two people are car shopping. One is a housewife with four children, and the other is a professional race car driver. Although both want the same type of product—a car—this is where much of the similarity between the two ends.

The housewife needs a station wagon to haul all the kids and groceries. Speed means very little to her, but comfort means a lot. She'll want air conditioning for those summer rides to the stores, a heater for the winter mornings, an automatic transmission, and a nice radio to drown out the screaming and bickering that often surrounds her.

The race car driver wouldn't even *consider* a station wagon or an automatic transmission. He's in the market for something small and sleek with a stick shift. In addition it must be lightning fast and maneuverable at those very high speeds. A radio, air conditioner, and heater mean virtually nothing to him, but comfort is a factor because some of his races are lengthy.

As you can see, the *needs* determine the type of vehicle best suited to each of these people. The same holds true for audiences of corporate programming, which is why program design is so important.

Program Design Information

Program design information is usually gathered in the early stages of script development, often at your first few meetings (see Figure 4.1, pages 32–33). It is acquired primarily from the client. Audience interviews and your producer are also good sources. Program design information has very little to do with the content of the program, and it is not conceptual or visual in nature. It is the raw material from which the design of your program is molded. It is information like

- the problem or need behind the program
- the program objectives
- an in-depth audience analysis
- the program's proposed utilization

We will examine a program needs analysis, one of the final written forms this design information can take, in Chapter 14. For now, though, let's look more closely at each of the areas just mentioned and see specifically how they could help establish the design of a script you might write.

The Problem or Need

This is the justification for making the program. It also acts as a basic conflict that the program objectives (we'll discuss these in a moment) are meant to resolve. A typical problem statement for a new program dealing with vehicle-backing procedures, for instance, might be written in a needs analysis like this:

> PROBLEM
> Employees who drive company vehicles do not understand safe backing procedures. As a result, $120,000 was lost last year in accident claims.

As you can see, this is both a major *problem* to the organization, and it sets up a basic *need*: the requirement to teach employees how to back their vehicles safely.

This problem (sometimes referred to as the *purpose*) should be one of the first topics of discussion for you, the client, and the producer. If, in your initial discussions, you begin to sense that a real problem or need does *not* exist, there's a good chance you shouldn't be making a videotape or film in the first place.

Objectives

These are a series of statements that, if accomplished, will help solve the established problem and thus meet the need. There are several types of objectives. The two most common are instructional objectives and motivational objectives.

Instructional objectives   To be most effective, instructional objectives should be stated in such a way that the viewer of the program could actually be *tested* on how well they were met. For instance:

PROJECT PROPOSAL AND TREATMENT

V-#______ Class______

DATE: __________

WORKING TITLE: __________

SUBJECT: __________

CLIENT: Name ______ Title ______ RC Phone ______

CONTENT EXPERTS/APPROVAL: Name ______ Title ______ RC Phone ______

Name ______ Title ______ RC Phone ______

Name ______ Title ______ RC Phone ______

Name ______ Title ______ RC Phone ______

Name ______ Title ______ RC Phone ______

DEPARTMENT(S): __________

DEADLINE: __________

PROGRAM PROPOSAL WRITTEN BY: __________

Problem:

(What is the problem this program is designed to correct? State succinctly.)

Background:

(What is the need for this program? What is the situation to date concerning the problem? What other solutions have been tried? What are the audience's prejudices towards the proposed program? What is the audience knowledge and/or experience with the subject?)

Figure 4.1 Project proposal and treatment form used by a large company. The form acts as a thought jogger for writers and producers in client meetings.

Objective
Having viewed the proposed videotape program, audience members will be able to state the three company-approved backing rules, as follows

1. Do not back up unless absolutely necessary.
2. Exit the vehicle and physically check the rear before backing.
3. Honk before backing the vehicle.

These objectives state very specific results we hope to elicit from our audience. They are sometimes called *behavioral* objectives because they set up a predictable behavior an audience member should be able to carry out, having seen our program. The key behavioral phrase in this objective is "be able to *state*." You can see that objectives like this make the effectiveness of a film or videotape program very measurable. If this is

```
Proposal/Treatment - cont'd
Page 2

Objectives:

After viewing this program the audience will be able to identify: (or use other
action verbs) Primary Objectives:  (Number objectives) Secondary Objectives:




Audience:

(Hourly?  Management?  General Public?  Describe)
Primary:     (workgroup/department)_____   Size_____
Secondary:   (workgroup/department)_____   Size_____




Distribution:

(How will the program be utilized?  How many copies should be made?  In what
environment?  Office?  C.O.?  Tailgate?  What print materials will accompany
the program?  Who is responsible for them?  Will program be evaluated?  If so,
how?)




Relationship to Company Goals:

(Using current company goals what are the benefits of the proposed program?
Describe and number.)




PROGRAM TREATMENT

(Include format  of program. Videotape, slides, multi-image? audio? why?
studio? EFP? actors #_____  Host #_____, talk show? documentary? Why? tape
format: 1", Betacam, 3/4")

(Treatment is a narrative description of what the program will look like and
it must include all content relating to the subject of the program.  Dialogue
may be used sparingly to highlight a point or two.  Excessive TV terms should
be avoided.  Write like a story and include who, what, when, where, why, and
how.)

Preliminary cost estimate _____
```

Figure 4.1 Continued

the agreed-upon objective and 90% of the audience members who view this program can actually state the three points, the program is a success.

By contrast, our objective might have read

OBJECTIVE
Having viewed this program, audience members will have greatly improved their backing ability.

This objective is very vague because of the phrase "will have *improved*." How would we know if the film or videotape actually accomplished this? What exactly is an improvement? Could we give a backing test to every employee in the company? At a great deal of expense, we probably could. Even if we did, however, and they all backed their vehicles wonderfully, how could we be sure the program actually "improved" their skills?

Although this type of semantic scrutiny may seem like splitting hairs, you will learn very quickly in the scriptwriting field that when it comes to instructional objectives *specific* agreement on what the program will accomplish is extremely important. This agreement should be made among you, the client, and producer. Like the problem and need information, it should be discussed early in the research process.

Motivational objectives   Motivational objectives can be stated less specifically because they do not require a *measurable* behavioral response from the audience. For example

OBJECTIVE
Having viewed the proposed program, audience members will feel very positive about the company's new product line.

Creating a positive feeling may be a very important objective of your program. If employees "feel very positive," however, we could not really test them in terms of any specific behavior. Feeling positive is a much more personal and general state that the employer hopes will manifest itself in future attitudes and the sales figures of the new product line.

Mixing objectives   Often a program will have both kinds of objectives. Perhaps you are attempting to do this:

OBJECTIVE
Having viewed the proposed program, audience members will

- be able to demonstrate the three sales techniques used to inform customers about the new Futura II
  1. Inquire with tact.
  2. Overcome objections.
  3. Confirm a sales call.
- feel positive about meeting this quarter's sales quota for the Futura II

In this case, the instructional objective requires that audience members be able to *demonstrate* a behavior. The motivational objective requires that they *feel good* about the prospect of selling the product. If both objectives were accomplished, the audience would be well equipped to kick off a very successful sales campaign on this product.

Audience Analysis

This is probably the most critical factor in the success of your script. To write an effective program, you must know exactly who will be expected to perform those objectives you've just established. You must also know a good deal about your audience. This audience analysis often falls into five general categories:

- size/discipline
- demographics
- attitudes
- needs
- knowledge level

*A*ll good writers care about their audiences: What do they already know about this topic? Do they care about it? Will this information help them do their jobs better? Will it make their lives easier? The successful writer uses this type of information to mold a message to which the audiences will be receptive.

—Patti Ryan
Corporation Producer

Audience Research: A Key   Case Study 4

An associate of mine once did audience research that uncovered some interesting facts and changed the direction of a program she was working on.

She was hired to write a script that introduced a new, more efficient set of working procedures in a manufacturing plant.

One of the questions she asked the client in her first meeting was what opinion most audience members would have of these new procedures. The client assured her that most people in the audience would feel very positive about changing over to the new procedures. There were bound to be a few cynics, he admitted, but these would be few and far between.

My friend was somewhat suspicious. She asked the client why he felt such a positive opinion would prevail when the procedures amounted to substantial changes in people's jobs—something they're usually not too crazy about.

The client's response seemed logical enough. There had been many changes during the past 2 years in the plant, and he knew for a fact that employees felt that current procedures were disorganized and chaotic. The new procedures, he reasoned, would represent order and structure. Thus they would elicit applause from employees, who felt they were long overdue.

In order to see firsthand how the old procedures had been carried out, the writer decided to visit the field and do a series of employee interviews. Although she had no questions on her list about how well received the new procedures would be, she found she was immediately inundated with negative flak.

Employees felt the new package amounted to more double-talk, a threat to existing jobs, increased paperwork, management tampering with a system that already worked fine, and executive ignorance of what really went on in the factory.

Having found this out, my friend's next task was how to break the news to her client. She went to the producer first and explained what she had discovered. As it turned out, she was lucky in two ways: the producer was excellent at working with clients, and in this case the client was not motivated by politics or ego. He simply wanted to get the job done right. When he was told what the situation was, he immediately did a quick survey of his own and verified what the writer had discovered. He then had her and the producer go forward with the script based on the new audience information. He also began implementation of other measures to help solve the internal problems.

Here are some examples of how an audience analysis might be written in a program needs analysis. As you read through them, ask yourself how they might influence the type of script you would write.

Audience size/discipline   This audience is made up of approximately 900 company truck drivers.

Audience demographics   Audience members are roughly 85% male. Most are between 30 and 50 years of age and have considerable seniority with the company. Approximately 2% of audience members have college degrees; 50% have high school diplomas or equivalent. Most audience members are from the lower end of the middle-income bracket.

Audience attitudes   Company truck drivers have overall positive attitudes toward the company, but they also feel they are experts in their craft and thus generally above any instruction on how to drive properly. Their initial attitudes toward a program on backing procedures will probably be somewhat negative or at best viewed as a waste of their time. In addition, most drivers will probably feel that following the company's new parking rules, which include exiting the truck to check the rear, will be time-consuming and unnecessary.

Audience needs   Initially, audience members will see little need for this program. They would argue that the real need is to educate warehouse employees on how to park their carts and place their containers so that they do not become unseen obstacles in large truck-parking ramp areas. In fact, when really nailed down, many audience members say that warehouse employees should simply not be allowed to park or leave containers anywhere near trucking areas.

Audience knowledge level   Audience members are very knowledgeable about driving and parking trucks. Most have been at their jobs for many years. What they will *not* be aware of is the newest data, which indicates that 99% of backing accidents could be avoided simply by exiting the vehicle and checking to the rear. They will also be unaware of the monetary impact backing accidents have had on the company and the equation that suggests that the time spent by drivers following the new rules will amount to far less than the current accident costs.

As you can see, many facts here would affect the design of a script.

For instance, if the audience is made up of middle-aged truck drivers who are 85% male, we probably wouldn't want a female narrator or hostess in this case.

If the drivers' *attitude* toward exiting the trucks is a main negative point, this part of the script deserves special consideration. It would probably not be a good idea to show a role-play in this area because actors agreeably going through the motions of doing the job according to an unpopular company rule would probably get a healthy round of chuckles and jeers. It would have little if any credibility with this audience. What might have credibility, however, is attention to two other factors: their lack of knowledge about how much backing accidents have cost and the

*T*he writer can represent the audience's interests to the client.

—Patti Ryan
Corporate Producer

equation that says checking is cheaper than paying the damages. These may not be popular ideas, but they're credible and they make sense.

Other factors in this audience analysis that would affect script design are the audience's

- education: Would audience members best identify with a studious or grass-roots approach? Probably a grass-roots angle would work better.
- age: Would young actors or interviewees be as effective as those in the age range of the audience? It's doubtful.
- social status: Would a story set in a rich, upper-class neighborhood work in this situation? Probably not.

The truck drivers' animosity toward warehouse employees, their seniority, and their job knowledge are among the additional factors the scriptwriter should consider.

Multiple audiences It is generally not a good idea to produce one program for two or more totally different types of audiences. For instance, you would not design a program on cost-control methods in the same way for both assembly-line technicians and top executives. They view and accomplish cost controls from very different perspectives.

There are times, however, when one program will be aimed at two *similar* types of audiences. As an example, you might produce one program on cost-control methods for technicians and their supervisors because both probably view the subject from similar perspectives. When this happens, it is often a good idea to designate one audience as primary and one as secondary. The primary audience then requires a complete audience analysis, and so does the secondary audience.

In the case just stated, the audience analysis might start out with the audience size and discipline as follows:

- primary: This audience is made up of 750 auto assembly-line technicians.
- secondary: This audience is made up of the 57 assembly supervisors who oversee the primary audience.

Each of the other audience categories would also be divided in this way. Objectives are also set up individually for multiple audiences. Thus, the client would still be able to predict accurately what the two different types of employees would gain from the program. For instance

OBJECTIVES
Having viewed the proposed program, audience members will be able to

- primary audience: state two ways they can control costs
  1. by following attendance policies
  2. by following assembly-line time factors
- secondary audience: state why attendance policies and time factors are critical cost savers:
  1. because when followed, these policies have been shown to increase productivity to maximum levels

Utilization

Utilization of the project you are working on—that is, how it will be shown—also influences the way the script should be written. Consider the following two utilization statements and their impact on your script.

1. This program will be utilized as a specific segment of a company course on safe driving. A company instructor will administer the course and will provide the introduction to the program as well as discussion before and after the program on the subject of backing. Students will also receive a course workbook that includes descriptive artwork and the new backing procedures.

2. Drivers will utilize this program on a stand-alone basis in the field. The program will be viewed in yard coffee rooms following the employee's shift. Each employee will be required to view the program and then pass the videotape on to another employee.

Would a program shown as part of a formal training course, with considerable support in terms of an instructor and workbooks, be written the same as one to be viewed by truckers at the end of their shifts on an honor basis? Absolutely not.

The first type of utilization would call for a program with a minimal number of facts because an instructor and the workbook would also be there to provide information. In fact this program might not even *need* an introduction, title, open, close, and the like. It might simply be a series of vignettes or interviews with authorities in the field expounding on the amount of insurance and material loss businesses must absorb because drivers don't exit their trucks before they back up. In this case, the instructor would provide the opening, the closing, and the discussion; written material and artwork would provide additional facts.

With the second type of utilization, the program had *better* include an open, and it had better be a catchy one. If drivers who don't really like the idea of the program in the first place are expected to watch and learn from it after a full day's work when they're hungry, tired, and eager to get home, something on the screen had better catch their attention quickly. Also, because no other type of support material will be provided, this version of the program must include a good deal more information.

Design Summary

As you can see, design research allows you to obtain a good deal of valuable program information. These basic "noncontent" facts, when combined with certain other factors we will discuss shortly, act as a solid program foundation that allows you to custom-design the most effective script for the audience you are addressing. As mentioned previously, this information should be obtained early in the script development process, beginning with the first client meeting.

With program design information in hand, your next step will be to explore the content, that is, the actual subject matter, you will write about.

■ ■ ■ ■ ■
You Script It 2

You are having a discussion with a client about a recently produced program. The client is unhappy with the final outcome of the show, and he blames this on your script.

You recall that the script was written for a show that was supposed to teach secretaries how to use a computer system. The system was designed for budgeting and word processing. The client claims the program failed to do what it was supposed to. "They were supposed to be able to use the system," he says to you, "but they can't."

"Can they turn on the system?" you ask.

"Yes, they can," he says.

"Can they word process?" you ask.

"Sort of," he says.

"Can they budget?"

"They can get into the budgeting program, but they really can't operate it very well," he says.

Later that day you are at home and you pull out your file for that project. You come across the program needs analysis. You turn to the objectives page and find two.

1. to assure that employees can use the Excet computer system to budget and word process
2. to motivate employees to trust and use the Excet computer system

What could you have done at the outset of this project that would have alleviated the client's current dissatisfaction?

5

□ □ □ □ □

Content Research

Content information, like design information, is critical to the success of your project. You cannot hope to write an effective script if you do not thoroughly understand the subject.

What's more likely to happen to the writer who does not do adequate content research is a very embarrassing script review meeting in which the client will begin to ask questions like the following for which the writer won't have answers.

- "Why was this part of the procedure included and not that part?"
- "You did realize that next year this changes to that, right?"
- "I'm curious why none of the statistics we provided you got into this section where they really count?"
- "This part is pretty general, isn't it?"

The writer will stutter and grope as it becomes more and more obvious that she simply didn't research the subject properly.

Another, even worse, possibility is that these inaccuracies will get by the client, producer, and content experts and actually be recorded on tape or film. When discovered after the fact, content errors can cost thousands of dollars to correct.

The single most important aspect of content research, then, is simply to be positive that you *thoroughly* understand the subject before trying to write about it. You achieve this important objective by acquiring the proper information and organizing it into a logical, easily understood structure.

The writer should do extensive research. This will allow her to head off and solve many communications problems before they ever get to the client or the producer.

— Tom Anderson
Public Affairs Director
Corporate Video Client

Acquiring Content Information

You will get most of your content information from clients, content experts, and audience members. Usually the exchange of this information happens in meetings. In each of these meetings, you would explore the five Ws and what I call H and B/D.

who
what
when
where
why
how
benefits/drawbacks

If you've studied journalism, the five Ws will be nothing new to you. They form the basis for most exploratory and factual writing. The last two items are equally important to your content research.

Although the subject matter and situation dictate your actual questions, let's briefly run through some of the typical *types* of questions you might ask in each of these areas.

Who

Who is implementing the project? Who is affected by it? Who will administer it? Who developed it? Who will be available to the audience to answer questions about it after the program is seen? Are these people experts on the subject? Are they qualified in the eyes of the audience? Are they popular or unpopular in the eyes of the audience? Where are they located? What are their phone numbers? Should their input be absolute or subject to verification? How available are they?

What

What exactly is your subject? How many parts or steps does it have? Is it complicated or simple? Can it be explained or demonstrated in one sitting, or must you go to a number of people and places to fit it all together? Is written, graphic, or other source material available on it? Where and when can you acquire it? Does it lend itself to visual presentation?

When

When will it take place? Once it takes place, will it be permanent or temporary? Will it take place at the same time for all audience members? If not, what is the implementation schedule? How many phases are involved? Is the schedule final or subject to change? How might it be changed? When might it be changed? Will audience members have advance notice?

Where

Where will it happen? Where *won't* it happen? Will it happen in the same place for all audience members? If there is more than one location, how many are there? If there is an implementation schedule, what is the order of the locations?

Why

Why is it being done? Are the reasons documented? Are the reasons economical? Are the reasons critical to the company? Are the reasons popular or unpopular with the audience? Is the reasoning long or short term?

How

How, exactly, does it work or happen? Are a number of steps involved? Can it be demonstrated? Is it complicated or simple? Is it done the same by all employees? If not, how do the steps differ? Why do the steps differ?

Make good friends of subject matter experts.

—Daniel Gilbertson
Corporate Scriptwriter

Benefits/Drawbacks

What positive things will it do for audience members? What negative things will it do? Will audience members recognize the benefits or drawbacks? Are they short or long term? Are the benefits or drawbacks of major or minor importance to the audience? Are they of major or minor concern to the company?

Interview Techniques and Research Sources

As we've established, interviews are often your primary means of acquiring both content and design information. The techniques you use to conduct those interviews may vary, but several general techniques are helpful in most situations.

Use a tape recorder if possible   Handwriting notes can be laborious and sketchy. Not only are you likely to miss facts but also you may find that later you have a difficult time piecing together the fragments of content that you did manage to write down.

Using a battery-operated portable tape recorder solves this problem. It frees you to think about what is being said and to converse with the client, content expert, or producer. It also allows you much more eye contact and interaction, which usually results in a warmer, more personal relationship. Later, you are free to replay parts of it or the entire meeting, including many "tonal" qualities you would have missed in your written notes.

The only time tape recorders should not be used is when they make the client uncomfortable. This happens occasionally when the subject being discussed or the frankness you've elicited from your client runs contrary to corporate policies or loyalties. As an example, if you've asked the client to tell you frankly why he feels his department's safety statistics are lower than some other department's, his truthful answer may be, "Because my boss won't let me provide any safety training." He would probably be wary of saying this, however, with a tape recorder sitting in front of him.

The solution is to ask all clients at the beginning of your meeting if they have any objections to a tape recording strictly for your own later use. Also mention that if they come to any part of the conversation they would rather not have taped, they need only gesture to you and you'll be glad to simply turn it off. Follow this up by reiterating that the tape is strictly for your private, personal use as an alternative to taking notes.

Use open-ended questions   Open-ended questions cannot be answered with a simple yes or no. An open-ended question forces a client or content expert to think about what she is saying. It also elicits an in-depth answer that often leads to other subjects or areas you may need to explore.

Some examples of open-ended questions, using our earlier example of the truck driver's parking program, might be:

- "Tell me about how employees park now."
- "I'm curious to hear more about how the $120,000 was actually lost."

- "What are employees' feelings about getting out of trucks to check to the rear?"
- "What are employees' feelings about the company in general?"

Those same questions, posed in a *closed* manner, would look like this.

- "Do employees park according to the current rules?"
- "The $120,000 was lost in backing accidents, then, right?"
- "I take it employees hate getting out of their trucks to check the rear?"
- "Do employees like working here in general?"

You can see that the former series of questions would result in much meatier answers. They would also lead to diversions that might be helpful to you. For instance, when the client is responding to the open-ended version of the question on what employees think about getting out of their trucks, she might say something like:

> Well, I think they feel it's basically a waste of their time. They all try to do a good job, and we're always emphasizing doing the job as quickly as possible, so they're usually in a hurry. They also happen to have very heavy routes at this time of year, so I guess their feelings are understandable.

As the client was saying this, you might have quickly noted two things: "hurry" and "heavy routes." Both seem to be legitimate reasons why the truck drivers would be against the new rule. As such, you might feel these points require more discussion later.

Guide the discussion   If you've prepared in advance for your meeting, you should have a good feel for what you want and even the order in which you'd like to get it. You could save some time, then, by guiding the discussion toward these areas. Again, you would do this by first taking brief notes as the client mentions those items and then by posing other open-ended questions like: "I'm not sure if I understand. . . ." or "Tell me more about. . . ."

The only danger in guiding a discussion like this is that you could focus it too closely on what *you're* after and restrict the client from bringing up some important issue. To alleviate this problem, be sure to ask the client at various times during your meeting if there is anything more she would like to add.

Strive for simplicity   Most clients and content experts know their subjects intimately. You will be hearing them for the first time. This can lead to the "fast-talking, fifteen facts over-the-head" syndrome. Your protective response is simple. If you don't understand something, be sure to say you don't understand. I have stopped many clients and in a very pleasant manner said:

> Excuse me. I know I may seem dense, but you just lost me completely. Let's go back through that slowly and very simply. I

Ask questions, even at the risk of sounding stupid. More often than not, the questions will not appear as dumb to the client as they do to you.

—Ken Gullekson
Corporate Scriptwriter

want you to "Dick and Jane" me through each step. Maybe we should even draw sketches as we go.

The result is usually a slower, simpler, much more understandable version of the topic.

Listen   This may seem obvious, but *not* listening is often the reason writers and other people walk away from conversations knowing only generalities about what they've just discussed. This may be all right for informal conversations but not for design or content research meetings.

Focus on what is being said. If you feel you want to respond or take a diversion, scratch a quick note on your pad and let the thought go. Refer to it later when the client or content expert has finished. Also, encourage the speaker to continue talking by nodding frequently and saying "aha," "I see," or "right, go on." Above all, try not to interrupt the client's train of thought.

Effective listening is a skill that all good corporate writers work at continually.

Tune in to the client's feelings   There are good and bad subjects, and there are good and bad times to discuss things. Try to read your client's temperament and adjust your style accordingly. Remember that the comfort factor mentioned earlier is one of the most important aspects of the client-writer relationship. You should always strive to maintain it. If you sense that your client is becoming uncomfortable for any reason, try to alleviate the problem. Take a break, perhaps, or change your line of questioning for the moment. If these attempts don't help, a frank question about what may be troubling him might do the trick. If all else fails, canceling the meeting and continuing it another time, perhaps with another person, may be the answer.

Other Sources

Content research will often take you out of the conference room and into the field. You may need to interview employees to get their feelings about controversial subjects. You may need to meet and talk to engineers who can describe why and how something is done. You may need to attend training courses in which your program will be used or on which it is being based. You may need to read company booklets, policies, or practices. You may need to make calls, watch other films or videotapes, see demonstrations, take pictures, or make sketches. You may even need to hop on a plane for a different state. The key is to pursue whatever it takes to make you an expert on the subject.

If your research will entail other than incidental expenses, the producer should reimburse you or advance you these monies in addition to your fee.

Organizing Content Information

The end result of all this factual acquisition could be a jumble of notes, tape recordings, copied pages, pictures, diagrams, tear sheets, and man-

When doing your field research, take along a small consumer camcorder, and shoot equipment, people, processes, and locations. It will not only provide you with good reference material, but it can also be of great help to the producer and director when planning the shoot.

—Alan C. Ross
Corporate Writer/Director

uals. Your next step is to organize this content information into a logical order. In other words, you must now develop a content outline.

The Content Outline

Content outlines take different forms. For some projects, a simple series of bulleted statements briefly outlining the facts will do. For others, a formally researched and written document is required.

As mentioned in earlier chapters, the type of outline you are asked to develop (if any) will be based on the amount and complexity of information involved in the project and, most likely, the client's or producer's preference.

Whether it is required of them or not, however, most writers find that developing some sort of outline makes the writing process smoother, faster, and more enjoyable. An outline provides a solid structure for what, up until then, has been a conglomerate of random information. This structure then becomes a kind of road map for development of the treatment and script.

Content Outline Formats

Informal The informal outline may be a logically structured series of sentence fragments like this:

sales decline—the basic problem
Sales are down since first quarter due to

- inactivity of sales force
- product lag time
- competitive insurance

OBJECTIVE
Return sales to 1 million annually by attacking three problems in first quarter:

1. fire up sales force with incentives
2. speed up product delivery
3. overpower competition with new promotional campaign

Sales force incentive program will include

- 35 trips to Hawaii
- $25,000 in bonuses
- executive dinners
- plaques

Whether handwritten on napkins, typed on a word processor, or spoken into a tape recorder, you can see that a script or treatment could be developed by following this outline point by point. The key is the logical arrangement of information, that is, placement of that information into an easy-to-follow structure.

During script development, the writer would use these statements as thought joggers. They would refer him to documents or other content sources in his possession for further elaboration. The incentive program

> *K*nowing what not to put into a script is just as important as knowing what to include. This is where many clients need the writer's and producer's guidance.
>
> —Gary Schlosser
> Corporate Executive Producer

facts, for instance, might be elaborated in a brochure or recent company bulletin.

Formal Most formal outlines are similar to this informal structure, but they provide a good deal more meat. After writing a formal outline, a writer might not need to refer to anything but that document to develop the project further. Most formal outlines have three major parts:

> **Introduction** information used to gain audience interest and present an overview of the primary facts, used as a *hook* or *tease*
>
> **Body** an in-depth exploration of the facts first brought out in the introduction
>
> **Conclusion** one or a series of concluding ideas drawn from the exploration of the facts in the body

As you can see, these three different parts are linked together by the content information itself. The introduction gains audience interest and briefly presents the main facts. The body expands on the specific details of the main facts, and the conclusion makes certain deductions about the main facts.

This method of organization—the order of presentation and the subtle repetition—makes the formal content outline an effective research tool.

Information organized in this manner is easily absorbed and retained by an audience. In essence, audience members are told what they will learn (the introduction); the learning process then takes place (the body); finally, they are reminded what has just been covered (the conclusion).

If you haven't already, you will no doubt eventually hear this reinforcement-based learning process referred to in this well-worn training phrase:

> Tell 'em you're gonna tell 'em.
> Tell 'em.
> Tell 'em you told 'em.

We will take a closer look at a complete version of a typical content outline in Chapter 14. To illustrate now, though, let's briefly look at three parts from what might be included in a typical formal content outline.

We will assume the topic is still our truck drivers' backing program. Part of a content outline introduction on this topic might read something like this.

> Because the accident costs are considerably more than those required to perform safe backing procedures, all company truck drivers should be trained on the backing procedures and be required to use them on a continuing basis.

The three steps involved in a safe backing procedure are: not backing the vehicle unless it is absolutely necessary; when backing is required, exiting the vehicle first and visibly checking to the rear; and, finally, after reentering the truck, briefly honking before actually backing up.

Studies have shown that performing this procedure takes only 22 seconds on average.

Later, the body of the same content outline would cover each of these three steps, as well as any other pertinent information, in much greater detail. The part of the body covering only the third step, for instance, might read like this.

3. After reentering the truck, briefly honk before actually backing up.

This step is required as a final caution to assure that no one has stepped behind or parked behind the truck after the driver has checked the rear. The step involves

- visually confirming that the truck is in neutral
- executing a brief, half-second blast of the horn
- waiting 2 to 3 more seconds while checking all rearview mirrors for any movement

Step three is important on all occasions, but it becomes most important when a period of time elapses between the time the driver first checked to the rear and the actual back up. This lapse of time often occurs when a radio communication interrupts the three-step backing process.

The conclusion of this same content outline might then refer to this same step.

Finally, step three becomes a measure of insurance. It guarantees that the backing path is still clear, even if the driver has taken the time to answer a radio request or has had some other distraction following the check to the rear.

As you can see, if this and the other content information were organized in this way, it would become a very clear, thorough, and logical means of presentation.

Like most other documents in this book, the content outline need not be according to any *exact* form. What's more important is that it be simply and logically structured so that the average reader can easily understand it.

If it were then presented to the client and producer, it would also provide an assurance that you had taken the time to fully explore the facts, that you have gained a very clear understanding of those facts, and that you are able to organize them in a way that will make them easily absorbed by your audience.

The spark of comprehension you display when you really understand a subject will give the client confidence in your ability to write about it.

—Ken Gullekson
Corporate Scriptwriter

■ ■ ■ ■ ■ ■

You Script It 3

You may one day find yourself in a corporate meeting in which the client seems dissatisfied with talk of objectives, audience, the problem, and audience analysis.

If so, she might suddenly stop the meeting and say something like, "Look, I'm not really an academic type. I believe in shooting from the hip and acting on instinct. I'd hoped we could talk about exciting program concepts today, instead of needs and objectives and all this school-type stuff!"

In such a case, what should you do or say?

Concept Thinking

Once the design and content information are clearly focused in the writer's mind, a creative thought process can take place (see Figure 6.1, page 50). Like a catalyst, this process will begin to combine the two types of information into something new: a concept.

Print Versus Visual (Showing Versus Telling)

Before getting into the process of actually developing concepts, we should take a brief detour to discuss the basic differences between writing for visual and print media. This distinction is important at this point because the medium for which you conceptualize may determine how you approach a subject. In other words, concepts developed for books or magazine articles will often be different than those for scripts. This is the case because writing for the print media differs from writing for the visual media in much the same way that telling differs from showing.

Telling (Print-Oriented Writing)

When you tell a listener something, you explain the subject with words. Based solely on those word choices, you attempt to coax the listener into imagining what you are trying to communicate. If you tell in verbal form, you are talking. If you tell in written form, you are writing for print. In this case, the words themselves are the only communication tool. They are the critical factor because they are what the recipient (the reader) will actually absorb to gain the information.

Telling-oriented concepts, then, are based on precise, often artistic word choices that are meant to spark a reader's imagination.

Showing (Visual Writing)

When expressing ideas in visual form, the writer does not concentrate so much on telling as on showing. Although the words remain important, their importance is primarily technical rather than artistic. Granted, they must communicate to the producer, director, and the client, but they will not be the final communication elements that spark a reader's imagination. Instead, the words in a script are simply a means of describing a series of sounds and motion pictures that will eventually communicate the message, not to a reader, but to a viewer.

You can see, then, that the critical issue for the scriptwriter becomes not, What words can I write on the page? but rather, What pictures can I create on the screen?

The ability to see a communication problem in these showing or visual terms is a key trait of all successful scriptwriters. They first use

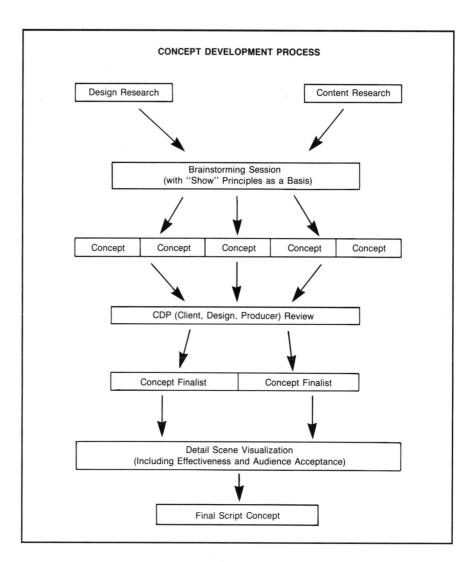

Figure 6.1 The concept development process. Although it may seem to be fluid or instantaneous, concept development usually evolves in a sequence something like this.

this ability when conceptualizing and then on a much more detailed level when visualizing or "shaping" those concepts into actual script scenes.

The Creative Concept

Keeping this basic "show" instead of "tell" idea in mind as we cover the next few chapters, we can now move forward into the processes of concept thinking and visualization. As a first step, we could describe the creative script concept as: *a creative visual idea based on all researched information that becomes the foundation for story line and visualization.*

Does this sound easy? Actually, it *is* a very simple process. Anyone can glance at a collection of information and come up with concepts all day long. The tough part is coming up with effective concepts: visual ideas that will communicate the proper content in a way that will help meet the objectives established to solve the communication problem that created the need for your program in the first place.

"Waking Up" a Phone Number   Case Study 5

A writer friend once found an interesting visual concept to wake up the end of a corporate commercial script for a company fitness center.

The spot was written very dynamically, suggesting hard rock music, sound effects, a constantly moving camera, slow motion, and rapid cutting. Although the writer knew the piece had to end with a slogan and a phone number for potential customers to call, he disliked the idea of character-generated words and numbers suddenly "stopping the show."

His solution was a creative use of sound and white characters on a black screen.

Our FRAME suddenly goes black. A portion of the slogan and telephone number, "DROP" in from "ABOVE" the FRAME. We HEAR a LOUD METAL CLANK, like a barbell dropping to the gym floor, as the title "HITS" the LOWER THIRD OF THE SCREEN. IT appears as follows:

gang get fit: call

The remainder of the message—the two ends—now SLIDE ON SCREEN from each side like weights being slid onto both ends of the barbell. As the ends of the slogan CLANK into place we HEAR the familiar SOUND this action makes. The final message reads

F. P. DORSE gang get fit: call 555-2245

Message HOLDS FIVE SECONDS, then, MUSIC STING and . . .

CUT OUT

Types of Concepts

The two basic types of concepts are: *program* and *segment*. Program concepts are general in nature. They are usually single unifying ideas that can provide an *overall* visual method of presenting a program. Some program concepts are:

Humorous role-play   Two cab drivers comically demonstrate the value of using seat belts on a whirlwind ride through the city.
Voice-over narrator with graphics   A voice-over narrator takes us through the year-end indices as we see animated graphic trend lines and artwork.
Host on camera with vignette segments   A host discusses substance abuse in an addiction recovery center as we see brief vignettes depicting employees becoming drug dependent.

People think in terms of similarities (this is similar to that) and differences (this is different from that). Use these common denominators as a foundation for understanding.

—Ken Gullekson
Corporate Scriptwriter

Segment concepts are the same types of visual ideas but focused on individual program *segments*. Some segment concepts (all of which might be used in the same program) are:

Graphics with digital video effects (DVE)   "Paint box" animation and DVE are used to provide visuals for the opening segment.

Split screen   The old and new products are seen side by side in a series of split screens to provide visual contrast.

Stills with character-generated titles   Still images of the test meter are combined with on-screen titles for "Part 1: The Meter Face—An Overview."

Concept Thinking

Both program and segment concepts are arrived at by means of a brainstorming and elimination process called concept thinking. It's worth our time to take a close look at a step-by-step analysis of how concept thinking takes place.

Remember as you read this, however, that concept thinking is primarily a *creative* process. It may not always happen exactly according to any linear, step-by-step method. In your case, some of the steps we're about to cover may not take place, at least not *consciously*. The entire process may seem to blend together into a fluid stream or sparks of ideas. Whatever the perception, however, concept thinking usually does involve a process that evolves something like this.

1. The content information drawn from research becomes the basis for a brainstorming session.
2. Brainstorming produces visual concepts for memorable ways to communicate that information via videotape or film.
3. These ideas are subjected to a client-design-producer (CDP) review that poses the question, Does this concept work for the client, the design factors, and the producer?
4. Based on the CDP review, many of the ideas are dropped. The few remaining ones are considered concept "finalists."
5. Detailed visualization takes place. Those final concepts that best develop into effective script scenes are the ones the writer chooses.

In order to follow this process to a functional result, let's look at how a series of program and segment concepts might be developed on paper.

The Concept Development Process

We could start by dividing a sheet of paper into three columns. The Content column on the left would be for information drawn from the content research, the Concept column in the center would be for the brainstormed ideas, and the right-hand column would be for our CDP review.

The writer offers ways to keep programs interesting and effective by balancing the needs of the client and producers with the needs of the audience.

—Patti Ryan
Corporate Producer

If we escaped to a quiet, fertile thinking environment, again re-turned to our truck drivers' backing program, and let the brainstorming process go to work, we might come up with a list that looked something like these:

Content	*Concept*	*CDP*
program	• host on camera in yard with role-play and graphic/title sequences	
	• testimonials—employee experts (not company) w/VO narration and montage open and close	
	• frank employee discussion around a table: completely natural, w/no scripted remarks—montage open and close	
	• role-play employee is first resistant, then almost runs over co-worker's child; decides to follow the rules	
3-step parking proc	• role-play—actor or employee	
	• use supered titles for each step	
	• animation	
only takes 22 seconds to check to rear	• clock in corner of screen as an employee checks	
	• employee testimonial	
	• host on camera	
	• cost comparison between checking rear and an accident	
checking is cheaper	• catchy, good title	
	• show actual figures on screen—charts, graphs, animation	
	• testimonial by company person who calculated it	
	• clock ticks off dollars for both ways—accidents and checking; checking comes up much cheaper	
	• split-screen, both happening at once	
can't restrict loc of warehouse	• truck jam @ pickup time to show confusion	
employees' carts, containers, cars	• employee en route w/containers/cart to show logic of route	
	• warehouse empl testimonial(s)	
open—grabber	• split-screen w/time and dollars ticking off	
	• sound bites debating both sides, showing some anger/resentment	

- sequence leading up to accident w/freeze frame at impact, then super title
- same as above but paying off the bills of accidents — fast-paced montage freeze on $ for title

As you can see, the initial result of this brainstorming session is a series of facts drawn from our content research information and corresponding groups of segment and program concepts that might work as visual ways to communicate these messages. Our next step would be to apply the CDP review. Once we had accomplished this, the list would probably look something like this.

Content	Concept	CDP
program	• host on camera in yard with role-play and graphics/title sequences	gd
	• testimonials: employees, experts (not company) natural/unscripted w/montage open and close	very credible
	• frank employee discussion around a table w/VO narration — montage open and close	most cred, hard to pull off
	• role-play — employee is first resistant, then almost runs over co-worker's child; decides to follow rules	no, too much to believe
3-step parking proc	• role-play — actor or employee	Okay, Empl better
	• use supered titles for each step	yes
	• animation	Expens(?)
	• host, VO role-play	yes, male driver type
only takes 22 seconds to check to rear	• clock in corner of screen as an employee checks	may be too condescending
	• employee testimonial	credible
	• host on camera	OK
	• cost comparison between checking rear and an accident	yes, if put right
checking is cheaper	• catchy, good title	too light?
	• show actual figures on screen: charts, graphs, animation	yes, if done right — not condescending

	• testimonial by company person who calculated it	maybe
	• clock ticks off dollars for both ways—accidents and checking; checking comes up much cheaper	catchy for visual imp
	• split-screen, both happening at once	good, but is accident too expens to shoot?
can't restrict loc of warehouse employees' carts, containers, cars	• truck jam @ pickup time to show confusion	good but lots of set up
	• employee en route w/containers, cart to show logic of route	okay
	• warehouse empl testimonial(s)	might anger drivers
open—grabber	• split-screen w/time and dollars ticking off	nice
	• sound bites debating both sides, showing some anger/resentment	good conflict okay (?)
	• sequence leading up to accident w/freeze frame at impact, then super title	gd impact
	• same as above but paying off the bills of accidents—fast-paced montage freeze on $ for title	gd

If you look over this list, you'll find that the CDP review brings out some interesting thoughts relating back to the design research, client, and producer factors we've been discussing.

For instance, in the "program" entry, which is first, a host on camera with role-play is considered good, but most credible to this audience is an employee roundtable discussion. The CDP brings up a producer problem with it, however; it is considered difficult to actually pull off.

The second content point is the three-step parking process, a key instructional objective. The CDP review here concludes that using a role-play to communicate this is acceptable, but actors will be less credible than real truck drivers with this audience. Animation is another possible concept, but the CDP review here assumes that it may be too expensive: a producer limitation.

In the next entry, the content point is the fact that the check-to-the-rear procedure takes only 22 seconds. The CDP review calls attention to the fact that using a clock to illustrate this may seem condescending to audience members who are very knowledgeable truck drivers. Testimonials from other truck drivers, however, are noted as very credible with this audience.

In the last entry, concepts for the program opening, one would show interview bites from both drivers and warehouse workers, each voicing his or her side of the issues. From an audience standpoint, this would create some engrossing opening conflict to draw in viewers. Nevertheless, the wise notation here questions whether showing that conflict is acceptable. This is a client consideration and a good one. Some clients insist on showing only the positive aspects of an issue, even if a little negative conflict would do an excellent job of communicating the point.

A quick scan of the rest of the sheet will confirm that the pattern continues: content on the left, concepts in the middle, and CDP concerns, which relate directly back to critical client, design, and producer factors, on the right.

When this exercise has weeded out most of the unworkable ideas, a new list including only the concept finalists might be developed. It would probably look something like this.

Content	Concept	CDP
program	• host on camera in yard with role-plays	gd
	• testimonials: employees, experts (not company) natural, unscripted w/montage open and close	credible gd
3-step parking proc	• role-play	employee
	• use supered titles for each step	yes
	• host, VO role-play	yes, male driver type
only takes 22 seconds to check to rear	• clock in corner of screen as an employee checks	may be too condescending
	• employee testimonial	credible
	• host on camera	yes
	• cost comparison between checking rear and an accident	yes, if put right
checking is cheaper	• show actual figures on screen: charts, graphs, animation; checking comes up much cheaper	yes, if done right — not condescending
can't restrict loc of warehouse employees' carts, containers, cars	• employees en route to show logic of stops/sequence	good
open — grabber	• sequence leading up to accident w/freeze frame at impact, then super title	gd impact

> • same as above but paying off the   gd
> bills of accidents—fast-paced
> montage, freeze on $ for title

As you read in our concept thinking steps, the last part of this process involves a transition of creative thinking skills. The act of conceptualizing begins to be replaced by detailed visualization. This step, in the end, provides the writer with the final concept decisions.

Before we move into visualization in Chapter 7, however, let's briefly look at a few other program concept ideas and discuss the types of content, design, production, and client information they were drawn from. As you read these accounts, look for the key CDP words and consider how they helped produce the final concept choices.

Other Examples

Host on Camera with Slides Transferred to Tape

A friend of mine once wrote a program for a client who wanted to train employees on the various functions an electronic calibration meter would perform. There was very little money available, and the client and producer agreed they needed something carefully structured but not flashy.

Because one objective of the program was to replace a supervisory presentation on the subject and because the audience was made up of blue-collar shop employees, the decision was made to go with an on-camera host in a shoplike setting similar to where the employees worked. The host, it was decided, would act as a peer-level instructor.

There was very little to see visually, however, and the nonhost segments were written calling for extreme close-up *still* shots of the meter's face. Superimposed titles were used as a means of support because the program objectives were primarily instructional. For instance, one part of the script read

ECU—RANGE DIAL. It is set at
Times 10 ($\times$10).

                                                      HOST VO
                                            However, when the range dial is
SUPER: $\times$10 = TEN TIMES   set at the ten times position, all
READING ON DIGITAL STRIP   readings are multiplied by 10 on
                                            the digital readout strip.

The result was a very simple, economically made, but effective program. The producer shot all the stills with a standard 35mm-film camera and then transferred these slides to videotape. These were intercut with the host footage and used with extensive supered titles. The program didn't make the most exciting use of the medium, granted, but a definite client need was successfully filled with a concept appropriate to the CDP factors at hand.

Music Video

The final awards banquet for a large sales force once called for the production of what turned out to be a music video. The objectives called for nothing but excitement, fun, and motivation. No information or instruction was required. The audience was made up of sales employees who had worked very hard during the campaign, and the client had a fairly hefty budget.

Because the theme for the entire 6-month campaign had been football, a Saturday football game was set up in a nearby park. Employees who had been involved in the campaign were invited to come and play touch football. They and their families were treated to Cokes and box lunches. A multicamera video crew also attended and videotaped the football game as well as the other festivities.

A motivational song lyric was also written, arranged, and produced. This musical sound track provided the basis for the final concept choice: a 3-minute, slow-motion montage featuring shots of all the employees who had been involved in the football game.

The completed program was played for these same employees 3 weeks later at the awards banquet. The viewing environment was a large auditorium. The lights were dimmed to signal the beginning of the meeting, and with no announcement the video suddenly played on a large-screen projection system. Seeing themselves, their family members, and peers playing football in slow motion to a very moving and motivational song drove the audience wild.

Documentary

Employees in a large company were suspicious of a new kind of decision-making process being introduced. Employee involvement (EI) was based on participative consensus decision making in no-holds-barred meetings between management and hourly employees.

Both groups of employees were wary of EI. Management employees felt their decision-making powers might be robbed by the system, and craft employees felt they wouldn't really get more than a token involvement in the real decisions made in the company.

To give the program audience credibility—a key objective—it was written as a documentary based entirely on interviews with employees who had seen the system work successfully. Because the client felt strongly that audience members should see parts of an actual EI meeting in progress, the interviews were intercut with unrehearsed shots from actual meetings.

The result was a scripted but seemingly unscripted, very direct program that was extremely effective at convincing employees and managers alike to give the system at least a try.

Children Interviewed on Clean Air

A program was required to convince employees in a large company to begin to do their part toward cleaning up air quality in the Los Angeles area. This meant, in particular, carpooling, riding buses and vans, and walking to work.

Role-Play Caution   Case Study 6

Some inexperienced scriptwriters lean toward role-play as a standard visual method for communicating most information. They tend to think that because scenes using actors are "Hollywood" or "visual" in nature, they're bound to be entertaining and thus effective. Experienced writers know that role-play can easily backfire in the wrong situations.

For instance, role-play is often *not* a good idea when you are communicating messages that are suspect or controversial for employees. Consider this message: "Working overtime and weekends without pay is good because it will help the company get back on its feet."

Although this statement may be valid, there will no doubt be quite a few employees who dispute the idea. To show them a role-play of an employee coming around to this line of thinking at the urging of a fellow model employee would probably produce more jeers and animosity than it would do good.

Suppose you are producing a program for a brand new sales force, however, and the message is: "Here are three surefire techniques for getting the customer to sign the contract in less than an hour."

Chances are this audience will be eagerly awaiting a role-play that will show them how to do this. If it's well written and performed, you'd probably be able to hear a pin drop between bits of dialogue.

This same principle holds true for other concept choices like humor or musical treatments. They can be very effective when used appropriately. The key, as we've discussed, is to base the script on CDP factors rather than on personal preferences or industry allure.

Both the client and the producer felt something memorable and motivational was called for because employees had seen several previous informational programs on the subject and, like most of us, they were procrastinating. Very little money was available to do the project.

After much consideration, the program concept of children talking extemporaneously about the effects of smog and their future in the city became the basis. These shots were intercut with three other elements: simple titles giving little-known harmful effects of smog on children, smoggy shots of cities and freeways, and colored drawings by the children portraying their ideas of what smog was like and what it did to humans and animals.

The children were interviewed on a playground to show them vigorously exercising out in the air. Their off-the-cuff remarks, mixed with simple but powerful facts and pictures on the screen, made for a moving and effective motivational program.

Concept Summary

In each of these cases, the main point to remember is that much thought was given to various program and segment concepts, and those decided upon were arrived at after a process like the one we've been exploring—

content information as the basis, brainstorming as the creative process, and finally applying the CDP factors—as a test of effectiveness.

As mentioned earlier, the conceptual process itself did not have to be executed in a step-by-step, perfectly organized fashion. The categories may have been written on several sheets of paper instead of in columns on one sheet. Part of the information may have been recorded on an audiotape and played back as the writer took notes. In fact, the information may not have been written down or documented at all. The entire process might have been a mental one. In whatever form it did happen, however, it used the types of information we've been discussing as a basis for an evolution of raw information into appropriate creative concepts.

Once this conceptual phase has been accomplished, the final step becomes choosing the actual concepts to be used from the few remaining finalists. At this point visualization takes over.

■ ■ ■ ■ ■
You Script It 4

During the preliminary client meeting on a project you've just been hired to write, the client and producer have provided you the following information.

Purpose of the Program
To convince employees a new automated service order processing system will not lead to layoffs and that it will improve their work environment greatly.

Background
The same system was installed 1 year ago in the company's plant in a different state. It has led to no layoffs, and the employees love it because of the many ways it has made their jobs easier and more pleasant.

Audience Attitudes
Audience members are very suspicious of the new system. This is partly because the company has undergone many organizational changes during the past few years, and employees are also insecure about the future of the company in general.

Audience Knowledge Level
Audience members know nothing of the new system other than what they have picked up from the rumor mill. Their current mode of operation is totally manual, and their general job knowledge is high.

OBJECTIVES
Having viewed the proposed videotape, audience members will

1. be able to name two ways the new system will improve their jobs
2. feel secure that the new system will not take away jobs
3. be motivated to give the new system a fair shake

With this information in mind, which of these three program concepts would probably be the best choice?

Role-Play
For a dramatic vignette in which a similar system has been installed, the program jumps forward in time 1 year and shows employees (actually actors) still in their same jobs enjoying the benefits of the new system.

Documentary

A factual program could be made up primarily of interviews with employees from the plant in the other state. They would openly discuss their own fears before the system was first installed and explain how it has made their jobs easier and resulted in no layoffs. The interviews would be totally unscripted.

On-Camera Host

A host discusses the benefits of the new system and actually shows the audience the new system in action. This is accomplished by shooting footage of the employees in the other plant utilizing the new system in their work. This footage is then intercut with the host segments, which would be shot in a studio. The company president would also be included in this program to give brief opening remarks assuring employees that the new system will mean no layoffs.

7

□ □ □ □ □

Visualization

*T*ell your story visually. *Write for the ear, but speak in images.*

—Daniel Gilbertson
Corporate Scriptwriter

As we have seen, ideas are communicated in various ways. In a typical conversation, they are exchanged verbally in the form of words, gestures, and inflections. On the printed page, they are read. In the sound media, they are heard. In the most powerful communication media of all, however, films and television, viewers actually *see* and *hear* those ideas created as "live" visual events before their eyes.

This sound and motion picture element makes television and films powerful and popular communication tools. This is also the reason the visualization process used to shape our previously developed concepts into actual script scenes is so important.

The Writer and Director

Visualization begins with writers. Based on the concept thinking they have done, they create a series of imagined motion pictures in their minds and describe them on paper according to a particular format. The first formal written description is the treatment; the second and final version is the script. A script, however, as we've noted, is not an end in itself, like a book. Rather it is an intermediate form of the writer's creation meant primarily to communicate a vision to another person, the director.

When the director takes over, it becomes his job to finish the creation the writer has begun. He too must visualize the script in fine detail. In doing so he is able to translate the writer's imagined motion pictures into live scenes that successfully convey the ideas the writer strove to communicate.

A Definition

We might define the scriptwriter's visualization process, then, as: *imagining a detailed series of events and audiovisual elements that transform creative visual concepts into potential script scenes.*

"Imagining," as you might guess, is the basis for the entire visualization process. We'll talk more about that in a few moments. The "events and audiovisual elements" are the same ones we discussed in our earlier analysis of a script: actors, locations, entrances, exits, sounds, titles, special effects, and so on. The "creative visual concepts" are the visual raw material we've just arrived at by brainstorming with our content, client, design, and producer information.

Print and Visual   Case Study 7

Writing for visual media differs from writing for print media in style and conceptual development.

	Print Writer	Scriptwriter
Style	writer to reader	technician (writer) to technician (director)
	word oriented	picture/sound oriented
	imagination driven	sensory driven (sight and sound)
Concept	pain due to loss of a loved one	pain due to loss of a loved one
Example	When John got home that night, he immediately thought of Jan. In his mind's eye he saw her long blond hair and slim, attractive face. As he sat on the couch, the pain of her loss began to overwhelm him.	John comes through the front door looking beat. As he drops his coat, he sees a picture of Jan on the TV. A look of pain comes over his face. He moves to the couch, sits down, and drops his head into his hands.

Both writer and director understand the need for the technician-to-technician writing style. They know that the images created based on the script replace the need for the print writer's in-depth physical and psychological descriptions. The director doesn't need these elements because she is reading the script not as she would a story or novel but rather as a transitional working document.

The Visualization Process

Just as with concept thinking, visualization appears to be a very simple matter. We would all agree that it doesn't take much thought to look at a basic concept and imagine that idea happening in real life. In the case of script development, however, two complications very similar to the ones discussed earlier become factors. First, the visualization process must produce detailed script scenes that will help solve the established objectives. Second and equally important, those scenes must elicit audience acceptance.

As you can see, regarding the visualization process in this way forces us to think of it not just in terms of how to visualize scenes but how to visualize effective scenes our audience will buy into.

With these thoughts in mind, let's start by looking more closely at one of the key elements in this process, the visualization process.

Visual "Mind Sight"

The ability to think visually seems to be easier for people who express themselves in terms of showing instead of telling. It is a process of "seeing" a scene play out vividly in the mind's eye. Here are a few practical ideas that can help that process happen.

1. Place yourself in a mood and environment that are conducive to the visualization process. For some people the mood might be quiet and contemplative; for others it might be an energetic, "wired" state. The environment for some might be a quiet room; for others it might be sitting in front of a typewriter or word processor; for still others it might be a rear booth in a fast-food restaurant or a bench in a crowded mall.
2. Write down what you visualize. This serves two purposes. First, writing down an imagined scene forces you to ask yourself just how clear the scene really is in your mind. If the detail has not been imagined, it can't be easily expressed on paper. Second, writing down what you "see" improves your ability to communicate your visualization in written form.
3. Give your visualization time to gain focus. It doesn't have to happen all at once. In fact, it rarely does. In more cases, it's a matter of visualizing and then letting those images brew for a time in the subconscious mind. Later, they tend to resurface, perhaps with some new angle, detail, or sense of heightened clarity. "Later" might be 5 minutes or several days.

Concept Visualization

With these visualization skills as a basis, let's return to several of the segment concepts we considered in the previous chapter.

One, you'll remember, dealt with the short amount of time it took the truck drivers to check to the rear of their trucks. That concept description and our CDP review of it looked like this.

Content	Concept	CDP
Only takes 22 seconds to check to rear	• clock in corner of screen as an employee checks	may be too condescending
	• employee testimonial	credible
	• host on camera	yes
	• cost comparison between checking rear and an accident	yes, if put right

In order to decide on the most effective and credible of the four concepts chosen, we need to do two things. First, we must use the visual thinking process just described to imagine each of them as they would play out in real life. This process includes writing them down. Second, that ever-critical audience acceptance factor must again become a part of the process. We must *always* consider our concept visualization from the audience's perspective and ask which works best.

As an example, we might visualize and then write the first concept: "Clock in corner of screen as employee checks" like this.

> We see a typical parking space in a company supply yard. There is the usual clutter around, such as containers, stacks of tiles, hoses, buckets, and a small carpet cart. The warehouse seems busy, with several people moving in and out of the large shipping doors and a few carts coming and going along the loading ramps. Several large truck-loading spaces are open. A semitruck approaches from around the corner and pulls to a spot in front of one space. The driver cannot see all of the area behind him from the cab. As his truck comes to a stop, a clock graphic with a second hand appears in the lower left corner of the screen. The driver's door opens. The clock begins to tick. The driver walks to the rear, checks out the situation, walks back to the front and enters the cab. When his cab door closes, the clock hand stops. The clock now comes forward and fills the screen. It is superimposed over the scene, which is now frozen. The hand has stopped on 21 seconds.

Now this seems to have been imagined clearly, and it would definitely make the point visually that less than 22 seconds is all it takes to check to the rear of the truck. From a visual standpoint, then, it is effective. The possible problem with it relates to our second criteria, *audience acceptance*.

If I happen to be a truck driver watching that scene play on a TV in a program that I know is meant to convince me to check to the rear of my truck, how might I react to it? I might say something like this to myself as the scene played out: "So, big deal. It takes twenty-two seconds. I could have told you that. It's still a pain in the butt!"

So we've visualized the scene well and written it down clearly, and it's strong on visual content. It's potentially flawed, however, in terms of audience credibility.

The next concept on our list was "employee testimonial," and our CDP review concluded it was "credible." Let's visualize this one.

> An actual driver is seen standing in a supply yard. In the background is a loading area with several trucks parked in it. The driver is looking at an off-camera interviewer, as might be the case in a news story or a documentary. In his own words, he says something like: "It's true. It really only does take 22 seconds. The point is it *seems* like more. It's always a pain in the butt when you have to unbuckle your seat belt, climb out of the cab, and walk back there. I can understand how not checking can be a problem, though. I've hit two containers myself and I really crunched a cart once. Luckily, there was nobody in it!"

Again the visualization is clear and well thought out. If I happen to be that same audience member and I see this concept play out on the TV screen, although it might not be quite as visually dynamic as a truck-parking scene, I would probably say something like, "He's right. It *is* a pain. It's true, though. We nail those containers all the time." So this concept may rate higher than the first one on credibility but lower on visual effectiveness.

The next concept, "host on camera," might be visualized this way.

> A host, who appears to be a trucker type, stands in a supply yard facing the camera. Behind him a truck is just parking. A driver gets out and starts checking to the rear. The host glances at the driver, turns back to the camera, and speaks to the viewers, saying, "Sure, it's a pain to have to do it, but it only takes about 20 seconds. And let's face it, we've all hit one too many containers for our own good." Behind the host, the driver now finishes checking, gets back into his truck, and backs in safely. The host exits the scene.

If I were the same viewer watching this visualization, I think I would probably find it fairly credible. However, I might be unconsciously thinking, "The guy who's talking *looks* like one of us and he *talks* like one of us, and what he's saying makes sense. It does only take a few minutes. He's still an *actor*, though. He's never really had to back a truck up himself."

So this concept probably comes up about medium on both the credibility and the visual effectiveness scales.

Now let's do the last one, "cost comparison."

> We have a standard split-screen showing a comparison. On the left is a still picture, a close-up of a truck bumper crunched into a small cart. Both are badly dented. Below this is a superimposed title: "Cost? $1870."
>
> On the right is a still frame of a driver standing at the rear of his truck checking before backing. Below him is another superimposed title: "Cost? $.12." Meanwhile a voice-over narrator says: "Multiply that accident figure by 64. That's how many there were last year. You'll get a yearly cost of about $120,000. And to check to the rear instead? That's right, only twelve cents per time, per employee. Even with *all* of us, that's only about $65,000 per year, one half the accident costs."

This is definitely a convincing way to visualize the contrast between the two costs, and I think I would find it credible as an audience member. There's a chance, however, that I'd also think, "It always comes back to money, doesn't it? What about people? What about *us*? What about the fact that we've got full routes every day, and we're trying to get them done in 8 hours?"

The Decision

What we have done is subjected four of our previous concepts to what we might call the "visualization test." We've imagined them actually happening and then written them down. In addition, we've placed ourselves in the viewer's chair and asked that all-important question, Based on what we know about this audience, will they buy it?

After going through this process, a decision must be made. Which visualized concepts will the writer present to the client and producer? The answer, of course, should be those that best communicate the message in visual form and are most credible to the audience.

Unfortunately, the ability to make these final concept decisions is not something that can be taught in this or any other book. Rarely is it a simple, black-and-white decision. There are countless ways to visualize nearly any scene, each with its strong and weak points. In the end, it comes down to a sense of what works on the script page and on the screen.

As for the examples we just explored, one possibility would be to drop the first and third concepts and go with a combination of the second and fourth. Although the employee interview is not as visually dynamic as, say, the first concept, a peer testimonial is probably the most credible way possible to put across a controversial idea to this audience.

We might then combine it with the fourth concept because the split-screen has visual impact and drives home the cost issue. Placing the two together might also diffuse any possible resentment about the emphasis on money. If the employee interview came right on the heels of the cost comparison, employees might never think to ask themselves that "people" question.

Variations

Just as with the concept development methods presented in this book, visualization methods vary. One writer might feel comfortable making notes about every concept she explores and then writing very detailed visual descriptions for each one. Another writer might take very few notes, doing most of the visualization work mentally. Still another might visualize and dictate the images into a tape recorder. Generally speaking, all you need is the proper mood, a place to think, and some way to document your thoughts. That can mean in your den with a word processor or at the local coffee shop with a pencil and a stack of napkins.

Remember also that your script may incorporate one program concept or ten segment concepts. Your decision may simply be a dramatic role-play, or you may decide on an on-camera narrator for the open and a continuing role-play throughout. Your concept may call for heavy artwork in the middle, a voice-over narrator throughout, and executive interviews interspersed with superimposed support titles. The number is unimportant as long as there is a sense of unity and continuity to the whole and it does the job effectively.

The Treatment

As I've mentioned, there are two documented results of the visualization process. The first is usually a treatment, which is a simple narrative description of the visualized concepts the script will present in their final forms. The purpose of the treatment is to describe these elements briefly to the producer and client, thereby gaining their approval to proceed to the script stage.

The key to good treatment writing is clear, simple descriptions of well-thought-out visualized concepts. No matter how clearly a writer "sees" his concept, if it's confusing or vague to the client or producer reading the treatment, it will most likely be unsuccessful.

I once saw a treatment, part of which read something like this.

PART I—SECTION 1:
INT SALES OFFICE

1. CAMERA "DISCOVERS" two sales representatives working diligently. WE DOLLY IN on one rep, MARTIN FOSTER, while he is talking on the telephone. We hear him say, "Yes, Mr. Johnson, that will be fine." The customer's voice is EQd as he responds to Foster, and we INTERCUT AS APPROPRIATE throughout this section.
2. When the conversation is completed, WE CUT TO the second rep, SUSAN DARWIN, in the FG. She has been watching Martin all along. Susan seems to be upset at Martin. An EXTREME CLOSE-UP REVEALS her fingers tapping on her desktop nervously. DISSOLVE TO:
3. INT MARTIN FOSTER HOUSE. Later that evening. . . .

A treatment like this would leave more questions in the client's mind than it answered. For instance, how does a camera "discover" something? What's a dolly? What do EQd and FG mean? What does *dissolve* mean? Just what is an INT? Why number everything, including the paragraphs?

Even if the client knew the answers to many of these questions, these little bits of script terminology are placed in the narrative description of this treatment like mental stumbling blocks that keep interrupting the flow of the written material.

This treatment would probably have been much better received had it been written in its entirety like this.

PART ONE: THE DISCOVERY
Two sales reps, Martin and Susan, are in their office working hard. Martin is on the phone with a client. As he says, "Mr. Johnson, that would be fine," we see that Susan is upset. She glares at Martin, tapping her fingernails on the desktop. Later than evening at Martin's house . . . .

This second version is simple and clear. That's because it's written in language we all understand, and the client is no longer forced to attempt to become a cameraperson or director in order to understand what he's being told. Keep these ideas in mind as you write your own treatments, and you are likely to please everyone involved with the project.

A Final Thought

Finally, remember that the visual aspect of your work is really what makes it unique among all other forms of writing. You should capitalize on this difference at every opportunity and make your visual concepts, treatments, and scripts continually impress the clients, producers, and directors who read them. In order to do this, you must hone your conceptual and visualization skills to a razor-sharp edge. Remember also that based on CDP factors some projects will call for exciting, highly innovative concepts that will allow you to challenge your concept thinking and visualization skills. Other projects will be more technical and perhaps require more traditional concepts. In every case, you should strive to produce the most effective writing you are capable of.

■ ■ ■ ■ ■ ■
You Script It 5

You are hired by J&J Concrete to write a script explaining new safety procedures and motivating employees to follow them. You obtain the following information in one of your initial client meetings.

Problem
Employees do not understand the company's safety philosophy. The accident rate is skyrocketing. A media tool is needed to communicate the philosophy to all employees.

Audience
Of the construction workers in J&J Concrete, 50% have high school educations, and 99% are males between 20 and 40 years old.

Audience Attitudes
Most audience members are proud of the fact that they are in a tough business and a macho philosophy prevails. They don't like beating around the bush, and they don't like excuses. Frankly, they don't particularly like safety procedures either; they feel they're for sissies.

OBJECTIVES
Having viewed the proposed program, audience members will

1. be able to state the company's new safety philosophy
2. be motivated to adopt the new safety philosophy as an inherent part of their jobs

Which of the following might be an effective program concept to pitch to the producer and client?

Role-Play
A J&J employee feels he's too tough for safety procedures. After arriving at work, he jokes about how silly they are to a co-worker who is applying them properly. Later

that day, the macho employee begins to operate a skip loader. He does not wear a hard hat, as the procedures require. He is hit on the head by a falling brick and sent to the hospital with a fractured skull. Luckily he is okay, but the call was close enough to make him a safety believer from that day forward.

Comedy

A slapstick series of mishaps befall an employee who forgets to follow the proper safety procedures: a hammer falls on his head knocking him out cold; he steps on a board full of rusty nails; he falls off a foundation into an excavation site; his toes get run over by a tractor.

Music Video

A rock band dressed as construction workers dances around a job site pointing out, through the lyrics, the dangers that await those who don't use the proper safety procedures.

Host Job Tour

A host who looks like one of the guys tours a job site explaining why the new safety procedures were developed and giving factual accounts of crippling or fatal accidents that happened to unsafe workers. The accidents are visualized with brief, dramatic vignettes. The host's attitude toward the audience is positive, but he is also basically matter-of-fact and unsympathetic. He says he knows that viewers are the kind of men who basically do what they want. He goes on to say that his goal is to pass on to them, before they make their individual choices, some true stories of what has happened to people like themselves who decided to ignore safety procedures. The accounts punctuate long-term family and personal suffering. The decision as to whether to follow the safety procedures is then left to the audience.

Script Formats

The motion picture script format is a very specialized form of writing that serves multiple purposes. First and foremost, it is designed to communicate the writer's visualization to the director accurately. This, as we have discussed, is to assure that the director will be able to reproduce the written scenes faithfully in live action.

Second, the script is an approval and budgeting tool for the producer. Regardless of all other documents that have been written on the project, not until the script is in her hands can the producer truly approve or disapprove the writer's concepts and visualization. Only in the script are the actual scenes and final dialogue or narration present as a complete communication entity. Moreover, only with the script in hand can the producer budget the production with accuracy. Although a treatment can tell him what types of locations and actors will be involved and provide a *general* idea of the extent of that involvement, only the script actually lays out those and all other aspects of the production in complete detail.

The script is also a "proof-of-the-pudding" approval document for the client. In no other document can she see the culmination of all research completely structured, with all dialogue, narration, visuals, and content in place.

For the other people in the production team, the script is also a specialized and very important document. For the assistant director or production assistant, it is formatted in a way that makes it easily broken down into numbered scenes, prop lists, shooting schedules, and so on. For the script supervisor, the script provides a standard format for notes that will offer the editor a directorial road map to the assembly of the different parts of the program.

Script Format Paradox

The importance of the script format can present a confusing paradox to the new writer because some new writers make the mistake of overemphasizing *format* to the extent of underemphasizing *content* or *concept*. They put so much effort into making sure each scene heading is exactly according to some stylebook that they lose sight of the fact that what's *in* that scene heading is what really counts. They worry more about the margins of the dialogue column than whether an actor can speak the words. They place so much attention on the special effects notations and transitional script terminology that they forget to ask themselves if those effects are really important to the story they are telling, or they become so concerned with including and capitalizing

Case Study 8   *Stock Footage Scripts: By the Numbers*

A producer I know came up with a requirement for three 2-minute employee self-help video segments to be produced with virtually no money. He decided to create them from stock footage from previously produced programs.

As the writer on the project, I was given "window dub" copies of the shows as sources for the visual side of my scripts. *Window dubs* are copies of the programs with time code numbers appearing in a rectangular area near the bottom of the screen. These numbers allowed me to select exact spots on the tapes to use as visual source material. For instance, in one case I picked a shot of a truck moving through freeway traffic that began at 01:05:10:00 (1 hour, 5 minutes, 10 seconds, and 0 frames) and ended at 01:05:20:15. As you can easily calculate, this piece was 10 seconds and 15 frames (a half second) long. I then wrote the proper narration to cover the footage.

This meant that the left side of these scripts had very little scene description. More important to the producer were the program titles and numbers, which told him exactly where that particular piece of footage was located. The shot above, for instance, was labeled

"Traffic Trials"
01:05:10:00 to 01:05:20:15 (10.5)
Company truck moving through freeway traffic

---

every camera angle that they fail to consider what that angle is doing to the visual continuity of the scene.

Seasoned writers have learned that although the script format is important it is also *flexible*. Above all, they know it should never override the need for clarity of expression. In essence, then, the basic script format should be considered a flexible communication tool with certain structural constants.

The three primary script constants are

Scene headings   indications of *where* and *when* the scene is taking place
Scene descriptions   brief but inclusive descriptions of the visual action that takes place, at times including sounds and suggested music
Dialogue/narration   any words spoken by either characters or a narrator

A brief example, including all three of these constants numbered accordingly, would look like this.

1.   INT OFFICE—DAY

2.   John enters the office carrying stacks of paperwork. He's

frazzled and mad. He plunks the paper down in front of Bill and vents his anger.

3.   JOHN
                    I have had it with you, Bill! This is the
                    fourth week in a row. When are you
                    going to get busy?!

Column and Screenplay Formats

The two primary types of script formats, both of which incorporate these three elements, are the two-column format and the screenplay format. Both are used in corporate television, but the two-column format is probably more common.

Other formats such as storyboards and documentary scripts are also used in corporate productions, but these are the exception rather than the rule. Because these formats use many common terms, we'll first examine the formats in general and in the next chapter cover the common terminology.

Two-Column Format

The two-column format divides the script elements into two vertical columns on the page. The video or picture part of the program is usually written in the left column, and the audio or sound part is written on the right.

The two-column format places the picture information conveniently beside the narration or dialogue it corresponds to. This format was developed primarily for multicamera television production, in which a director must switch signals from several cameras "on the fly" with verbal cues. These switches are called *take points* because the director says, "Take 3," "Take 1," or whichever camera or source she has decided on.

The two-column format works well in this type of directing because the take points can be easily marked off in segments with the action and the director's notes beside the narration or dialogue. Other information such as superimposed title notations and special transition marks can also be written in conveniently beside the words that cue them to happen.

A typical two-column format script for a live television production looks like the example in Figure 8.1, page 74. As you can see, the three constants—scene heading, scene description, and dialogue/narration—are present but separated into two vertical columns on the page.

When a director actually uses a script like this in a multicamera production, she marks it with take points, camera numbers, and other notes. It ends up looking something like the example in Figure 8.2, page 75.

Aside from live productions, the handy placement of pictures beside words has made the two-column format a predominant and very effective style in the corporate world for video, film, and slide programs.

```
FADE IN:

ROLL IN #1 - Graphic countdown  (SOT)
with DISSOLVE to MAIN TITLE
ANIMATION:

     CENTRATEL:
   Power Selling!

DISSOLVE

INT. STUDIO

In this ESTABLISHING SHOT
we see all members of the
CentraTel sales panel:  BILL
DAVIS, JOHN PORTEN, MARY MILES
and DAVE DORMON.  They are seated
at a conference table having
a round table discussion.  As
Porten turns to CAMERA we...      (MUSIC UNDER AND OUT)

DISSOLVE

ON PORTEN                              PORTEN
                               Hello, and good morning.  I'm
SUPER PORTEN'S TITLE           John Porten, and I'm happy and
                               excited to be your host for
                               today's broadcast.   As I'm
                               sure you already know, we'll
                               be talking about CentraTel,
                               an important new product with
                               exciting possibilities for
                               the future.  But before we
                               get started, let me introduce
                               the members of our panel.
                               On my left is
ON DAVIS - SUPER TITLE         Bill Davis from product
                               marketing...
ON MILLS - SUPER TITLE         Beside her is Mary Mills, our
                               resident advertising expert...
```

Figure 8.1 A two-column script page from a live corporate broadcast.

Screenplay Format

The screenplay format is the original Hollywood script format. It has traditionally been used to shoot film or single-camera productions, which often take place on location instead of in a studio.

In single-camera productions, the director is not working on the fly with multiple sources. Instead, one camera is moved from position to position, and the same action is repeated several times to create "coverage" of the scene. When edited together later, pieces of the scene from the various camera positions appear to be different perspectives of the same event.

The screenplay format does not split the sound and picture into columns. The scene descriptions are written across the entire page preceding the narration or dialogue associated with them. The dialogue or narration is then written in a wide column down the center of the page.

A typical screenplay format script looks like the example in Figure 8.3, page 76. Again, we have the three basic constants, but, as you can see, this format would lend itself less to the multicamera, take point style of directing. It would be perfectly acceptable, however, for a single-

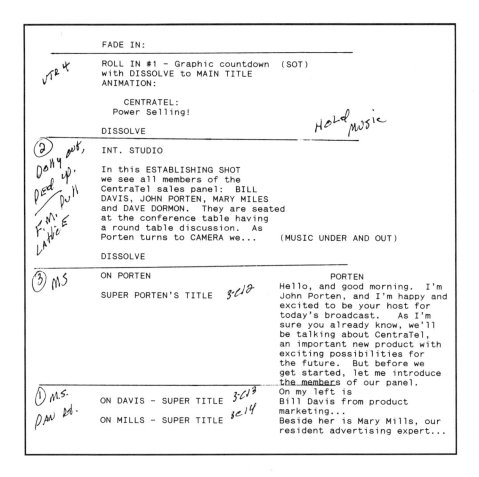

Figure 8.2 Script page from Figure 8.1 after director has marked it for production.

camera breakdown. Once that had been accomplished, it would look something like Figure 8.4, page 77.

Storyboard

The storyboard format is normally used for short productions such as commercials (see Figures 8.5 through 8.7, pages 78–81). Its intent is to show the client and producer artwork or photographs of the intended visualization in relation to the dialogue or narration. In order to accomplish this, the storyboard is usually made up of a series of pictures or frames rendered by the writer, director, or an artist. Each frame suggests what would be seen on the screen, and corresponding narration, dialogue, or scene description information is shown beneath or beside it. In larger productions, storyboards are sometimes used to graphically visualize certain important *parts* of a script, such as hard-to-describe graphic sequences. Directors often develop simple storyboards as a personal visualization tool during the preproduction planning stages.

Because the screenplay and two-column format are both common in the corporate world, a writer should be able to work in either style at his producer's request. Storyboards may not be as common, but the writer should also be familiar with this type of script presentation, should the need arise.

```
                              FADE IN:

EXT. INSTALLATION YARD - DAY

It is 7:45 AM.  Trucks are getting ready to roll.  Tailgate
meetings are just breaking up.  Two installers, DARRELL and
JILL are talking beside their trucks as they straighten up
some last minute orders.  Jill chuckles, turns to John, and
says...

                         JILL
              Ready for today's lesson in
              politics?

                         DARRELL
              Sure.

                         JILL
              How do you know when an
              installation field analyst dies?

                         DARRELL
              I don't know.  How?

                         JILL
              The donut falls out of his
              hand!

Both installers roar with laughter.

INT. FIELD ANALYSIS OFFICE - EVENING

A meeting in progress.  A group of about six field analysts
are gathered around a table.  Sleeves are rolled up, collars
lose.  Crinkled coffee cups and stack of papers are strewn
around the conference table.  There is lots of tension in the
room -- pressure, urgency.  ALBERT DALE, the field manager
step to the front of the group.  He pauses, looks the group
dead in the eye, and says...

                         MANGER
              All reports are cancelled.
              We all hit the field tomorrow
              at six AM.  I want to know
              exactly what we need to do out
              there... and fast...
```

Figure 8.3 A screenplay format script page.

Narration Script

A narration script contains only the spoken words. It can be written either in the column format with no scene descriptions in the left column or as a single narration column down the center of a sheet of paper.

The narration script is often used for executive "talking head" video shoots in which someone like the CEO decides to present a videotaped message to employees. Narration scripts may also be required for audio projects (usually audio training cassettes) or situations in which the producer has predetermined the visual part of the script. Audio projects may also incorporate *dialogue* instead of narration.

Although the narration script has no visuals, it is a common requirement in the corporate world. Executives often want to appear on camera; when they do, they need scripts. For this reason alone, the corporate writer should be prepared to write a narration script whenever it is requested.

The following is a short, partial example.

```
                                    FADE IN:

⟨101⟩   EXT. INSTALLATION YARD - DAY
-Est. High
-Go to Low       It is 7:45 AM.  Trucks are getting ready to roll.  Tailgate
 angle +         meetings are just breaking up.  DARRELL and JILL, two
push to          installers, are talking beside their trucks as they straighten
                 up some last minute equipment and service orders.  Jill
                 chuckles, turns to John, and says...
2 shot
+                              JILL
singles          Ready for today's lesson in
                 politics?

                               DARRELL
                 Sure.

                               JILL
                 How do you know when an
                 installation field analyst dies?

                               DARRELL
                 I don't know.  How?

                               JILL
                 The donut falls out of his
                 hand!

To W-S-          Both installers roar with laughter.

⟨102⟩   INT. FIELD ANALYSIS OFFICE - NIGHT
Pull Bck
over table?      A meeting in progress.  A group of about six field analysts
And Hold?        are gathered around a table.  Sleeves are rolled up, collars
w/ fast cut      lose.  Crinkled coffee cups and stacks of papers are strewn
reactions        around the conference table.  There is lots of tension in the
                 room -- pressure, urgency  ALBERT DALE, the field manager step
                 to the front of the group.  He pauses, looks the group dead in
                 the eye and says...
singles
MS/ CU                         MANGER
                 All reports are cancelled.
                 We all hit the field tomorrow
                 at six AM.  I want to know
                 exactly what we need to do out
                 there... and fast...
```

Figure 8.4 Script page from Figure 8.3 after director's production marks and notes.

HARROLD AVRELL NARRATION SCRIPT
Opening Sequence for "Today and Tomorrow"

MR. AVRELL

Before you view the following program, I felt it was important that you first understand the general direction our company will be taking in the coming decade. In addition, you should be aware of how our corporate mission statement has been focused on helping guide that direction. The decade ahead will bring exciting challenges and difficult times.

The Documentary Script

Not all programs are scripted *before* they are shot. In documentary film-making, a subject may be explored with the camera as a first step, and

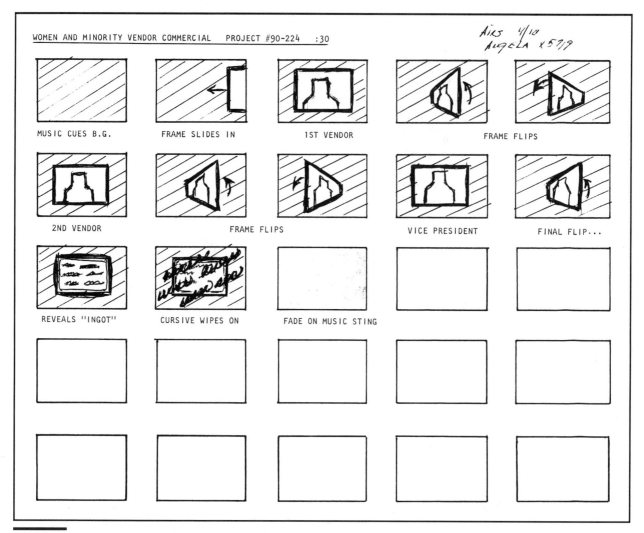

WOMEN AND MINORITY VENDOR COMMERCIAL   PROJECT #90-224   :30

Airs 4/10
Angela x5719

MUSIC CUES B.G.   FRAME SLIDES IN   1ST VENDOR   FRAME FLIPS

2ND VENDOR   FRAME FLIPS   VICE PRESIDENT   FINAL FLIP...

REVEALS "INGOT"   CURSIVE WIPES ON   FADE ON MUSIC STING

Figure 8.5 A simple storyboard created by a producer. This example was used to visualize a graphic sequence that was the basis for a 30-second commercial. For the actual script, see Figure 8.6.

script wraparounds and transitions may be written later to help place the visuals in a proper structure and provide only the necessary support narration.

When this is the case, an outline type of script is sometimes needed to help guide the director or cameraperson during production. The documentary script serves this purpose. It may suggest general visual coverage and provide content points for discussion by people being interviewed as part of the program. Scripts written for documentaries *after* production are often simply narration scripts. This is because the visuals will have been previously recorded and no doubt already decided on.

The following is a documentary outline script used for a program on air quality. The program was intended to explore the air-quality problem from the point of view of children.

```
WMBE/ALLISON  :30  90-224, 3/12/91

                              FADE IN:

An upbeat MUSIC SELECTION begins.  At the same moment, the
black screen turns to a colorful BACKGROUND DESIGN.  As
this background is established, a GRAPHIC "MARBLE" FRAME
SLIDES ONTO THE SCREEN.  Inside the frame is a FROZEN IMAGE
of a selected GTL W/MBE VENDOR, seen in her own business
environment.

As the frame reaches center screen, the vendor's image
UNFREEZES.  She is facing an off-camera interviewer who is
not seen.  She tells the interviewer (in unscripted,
off-the-cuff fashion) something like:

                         VENDOR #1
               I've worked with GTL several times.
               They're sincere about their commitment
               to women and minority businesses --
               and they're very supportive.

The frame containing this person's image now FLIPS OVER ON
SCREEN.  On its reverse side is a SECOND VENDOR.  He says:

                         VENDOR #2
               It's no free ride.  You have to know
               your stuff.  But GTL is willing to give
               you the shot at proving that.

The frame now FLIPS AGAIN.  On its reverse side is an image
of HUMAN RESOURCES V.P., MIKE ARBEL.  He says:

                         MR. ARBEL
               We don't give opportunities to people
               just because they're women or minorities.
               We give opportunities to people because
               they're good at what they do.

The frame now FLIPS AGAIN.  In doing so, it becomes an
INGOT LIKE PLATE EMBOSSED WITH THE WORDS:  "GTL Women and
Minority Business Partnerships"

OVERLAID on the plate, the following words now appear,
diagonally, in brightly colored, cursive letters:  "Doing
Business With the Best!"

As the MUSIC CULMINATES AND GOES OUT ON A STING, we...

                             FADE OUT:
```

Figure 8.6 Script that provided the basis for the storyboard in Figure 8.5.

WHO SOILED THE RAINBOW?
A Clean Air Documentary Program

1. Possibly open with coverage of kids in a school playground. Intercut with shots of dense, hot city smog, freeway traffic, newscasters talking about the deterioration of the ozone layer, and so on.

2. Sixth-grade kids interviewed on school ground. No scripted answers. Possible content points:
    *   Hot days are the worst. Sometimes their eyes hurt bad.
    *   Moms and dads may not know it, but kids get it worse because they play outside all day, especially in the summer.
    *   Sometimes, their chests hurt. It's scary.

Figure 8.7 Storyboard
developed to illustrate a
director's suggested visualization
of a short sales commercial.

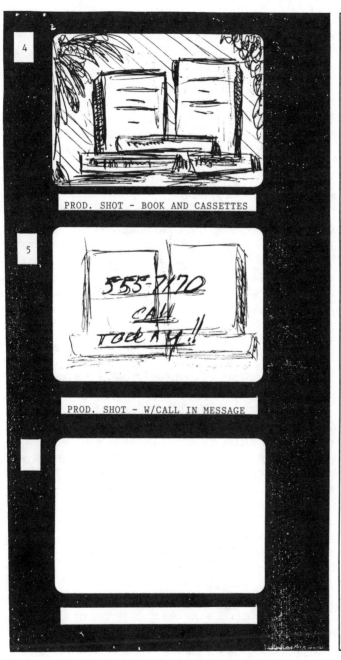

PROD. SHOT - BOOK AND CASSETTES

PROD. SHOT - W/CALL IN MESSAGE

Audio

If you'd like to order one of Tom Artten's books, or his complete video and audio cassette sales course,

Call now!

Case Study 9   *Executive Test: Know Your Subjects*

I was once asked to write a brief, "screen test" narration script for a new key executive. According to the client, its only purpose would be to let the executive and the public affairs department see how he photographed on videotape. Any coaching help he might then need before appearing in company broadcasts would subsequently be arranged for him.

When the producer and I discussed the project, however, she asked that I do a little research on the executive's management style and the things he considered important issues. She reasoned that if the script were centered around these issues, the executive's performance would thus be a more accurate indication of his true screen presence.

I agreed wholeheartedly and wrote a 3-minute narration script based on information I was able to find out about the executive's management style and operational priorities.

I was later told that when he showed up and began reading the script in the studio, an executive vice president who had accompanied him made the comment that the material he was presenting sounded accurate and very sincere. The taping that day was later included, with a narrator's introduction, in the next issue of the company's quarterly news program. It turned out to be a perfect way to introduce the executive.

Although this doesn't describe any scene *specifically* as it might happen in front of the camera, it does give the director a good general direction toward which to steer his visual coverage and interview segments. Later, the best of those shots and interview segments would be edited together, and narration would be written to supplement them.

Documentary scripts like this are also sometimes mixed in as segments of, for instance, a host on-camera script. This might be the case when portions of a program must be pre-scripted for instructional or motivational purposes, but the program is also heavily dependent on unscripted interview sequences.

The following example is from a script outlining employee assistance services available to employees in a large corporation.

INT MARCI ALBERT'S OFFICE—DAY
As we go to a SHOT of Marci in a counseling session, we first hear the host finish his statement.

                     HOST (VO)
            . . . so let's start with Marci Albert,
            EAP administrator for the Northern
            Area. Her degree is in psychology, and
            her love is watching people become
            happy and productive members of the
            team. . . .

EXT MARCI'S OFFICE—VERANDA—DAY
MARCI—Now being interviewed

SCRIPT FORMAT SUMMARY

Screenplay
Scene descriptions across full page, narration or dialogue in center column.
Original "Hollywood" Style.
Works well for single camera productions.
Feature films and teleplays scripted in this format.
Used extensively in corporate media.

Two Column
Two columns: Scene descriptions on left, narration and/or dialogue on right.
Used extensively in corporate media.
Works well for multicamera productions.
Also used for slide shows.

Documentary
General outline format, incorporating "bullet" statements and suggested questions for
    interviewees.
Acts as a general guideline for director.
Used in corporate media, often as a part of a script.

Story Board
Used extensively in commercials.
Used to illustrate a visual concept.
Often used in corporate media for short projects and illustrations of segment concepts.

Narration
No visuals.
Screenplay or two column style.
Usually for "talking head" executive presentations.

Figure 8.8 Script format summary.

                    MARCI
              (In her own words)
    1. her approach to the EAP philosophy
    2. how she prefers to deal with alcohol and drug abuse victims
    3. all EAP services available to employees
    As she completes these thoughts, we . . .
                                          DISSOLVE

Finally

Now that we've had a chance to examine the standard packages that much of the common script terminology goes into, let's have a look at the terminology itself.

■ ■ ■ ■ ■

You Script It 6

How might the following section of a program treatment be rewritten to better serve its purpose?

1. INT JOHN'S OFFICE
    CAMERA FOLLOWS John in with Mary close by. The two are discussing ARNOLD'S recent performance appraisal. We see a CLOSE-UP of the rating sheet as it is discussed. Arnold now enters. In a CLOSE-UP we see that he is upset over what he has just heard John and Mary discussing.

2. We CUT TO Arnold's home, 2 days later. CAMERA FOLLOWS Arnold's WIFE, JENNIFER, from the kitchen to the dining room table with a large roast. An exchange now takes place in which the dialogue between Arnold and Jennifer REVEALS how upset Arnold is over his poor rating. Suddenly, he REVEALS he is actually thinking of quitting.

9

□ □ □ □ □

Script Terminology

Like the three primary constants discussed in the previous chapter, script terminology is common to nearly all scripts, regardless of format. This is because it communicates important information to the production team in a language they use daily.

Script terminology can be broken down into four general categories.

editing terms
camera terms
scene heading terms
sound terms

Editing Terms

These terms note how a program will be edited. Like most other technical notations, they are usually capitalized in the script. This is done to make them stand out clearly for the director, producer, and editor.

Fade In/Fade Out

FADE IN and FADE OUT are used to note the open and close of a program or a major transition within the body of the show. They mean "fade the picture in from black" or "fade the picture out to black," respectively. These two editing cues are sometimes written as: FADE UP and FADE DOWN or UP FROM BLACK and FADE TO BLACK. All versions mean the same thing.

The black, by the way, can also be a different color. A script may call for a scene to FADE TO WHITE, for instance, for some dramatic or visual effect.

A typical FADE IN would be used this way:

FADE IN

INT OFFICE—DAY. JOHN MITCHELL is just pouring his first cup of coffee of the day.

Dissolve

This is a term meaning "dissolve one picture into another." In reality, a DISSOLVE is two fades, one picture fading in while the other picture is fading out. When this happens on screen, it causes an overlapping visual effect. A DISSOLVE is also a major transition second only to a FADE. It is meant to suggest a passage of time or a major change of location. For instance:

```
INT OFFICE—DAY
Mel and Adrian both peek around the corner and suddenly spy
each other. They walk forward, shake each other's hands, and,
just as in their school days, both begin to squeeze as hard as they
can.
                                                      DISSOLVE

INT DOCTOR'S OFFICE—DAY
Mel has lost again. He is in the doctor's office getting a cast put
on his broken hand.
```

The danger with DISSOLVES and sometimes FADES is overuse. As a writer you should keep in mind that a DISSOLVE is actually a special effect. It requires extra time in the editing process and can sometimes lead to costly editing delays.

As always, let necessity be your guide. Don't write in a DISSOLVE if it's not necessary. Instead, you should visualize both scenes in a particular transition; if the DISSOLVE is not needed, simply go with a CUT instead.

Cut

A CUT is the most direct and frequently used method of getting from one scene or shot to another. A CUT happens instantaneously. It simply means one picture ends and another begins.

Although many writers include the word CUT or CUT TO between nearly all scenes, in most cases this is not necessary. If one scene stops and another starts, the automatic assumption on the part of the producer, director, and editor will be that the transition is a CUT.

For instance, this transition

```
INT JOHN'S HOUSE
John picks up the phone and dials.
                                                      CUT TO

INT DAN'S HOUSE
The phone on the kitchen counter rings.
```

could just as effectively be written like this.

```
INT JOHN'S HOUSE
John picks up the phone and dials.

INT DAN'S HOUSE
The phone on the kitchen counter rings.
```

Wipe

A WIPE usually means one picture is wiped off the screen while another picture is revealed behind it. WIPES, like DISSOLVES, suggest a major change of time or location. There are many different types of WIPES, however. CIRCLE WIPES form a shrinking or expanding circle usually from the center of the screen. One picture is lost in the center of the circle and another is revealed around it. BOX WIPES perform the same effect in a box shape, CLOCK WIPES do it like a rotating clock hand, and there are many others.

A CIRCLE WIPE might be used this way.

```
EXT OFFICE—DAY
John exits the building and gets into his car. As he drives off into
the dead center of the screen . . . .
                                        CIRCLE WIPE INTO
```

Like DISSOLVES, WIPES are special effects requiring extra time and complexity to perform in editing. They, too, should therefore be used sparingly.

Digital Video Effect (DVE)

A DVE notation normally precedes the request for something like a picture that shrinks to a spot and flies away, splits into fragments, or warps into a cone and zips off the screen. There are numerous DVE "moves," as they are called, and like WIPES and DISSOLVES they are special effects. DVE moves, however, are even more costly to perform because they require digital effects generators that cost hundreds of dollars per hour to rent. An example of a sequence of DVE script notations might be written this way:

```
                                        DVE—PAGE TURN TO
ESTABLISHING SHOT—a desert landscape.
                                        DVE—PAGE TURN TO
ESTABLISHING SHOT—An expanse of snowy mountains.
                                        DVE—PAGE TURN TO
ESTABLISHING SHOT—A tropical coastline.
```

Camera Terms

Camera terms are usually also capitalized in the script. They let the director and camera operator know what type of focal length, camera placement, or movement you had in mind when visualizing the scene.

In order to keep your script as simple and effective as possible, camera terms should be kept to a minimum. Because you will still need to know and use this terminology, however, the following explanations should clarify those used most often.

Wide Shot (WS)

This is a camera focal length providing a wide angle of the scene. WIDE SHOTS are usually used to establish locations or to accommodate a scene in which action is happening in a large area.

WS—OFFICE. Bill and Dana are seated to the left at their computers. Ellen enters from the right and sneaks up behind them.

Establishing Shot

This is usually a wide and often distant shot used to establish a *major* location for the first time. Many times it is an exterior. For instance

EXT COMPANY HEADQUARTERS—DAY
ESTABLISHING SHOT of the front door and parking area.

Long Shot

This is a shot made with a long focal length. LONG SHOTS actually magnify the scene, thus things can appear disproportionately large in the camera's frame. You might use a long shot this way.

LONG SHOT—SEDAN. The car moves down the highway approaching us from a distance. Heat waves ripple up off the hundred-degree tar in front of it. Bill is behind the wheel, wiping his forehead with a towel.

Medium Shot (MS)

This is usually a shot of one individual from about the thighs up. It is sometimes more generally referred to as a SINGLE or ANGLE ON, or sometimes simply as ON. A MEDIUM SHOT could be used following a WIDE SHOT to begin to move in on the action. It might also be used when the actor should be seen with a small part of his environment included in the shot.

MEDIUM SHOT—BILL. He gets up from his chair, walks to the corner, and stands beside the computer and printer setup. He tears out a sheet of paper.

ANGLE ON DALE. He gets into the car.

ON GARY. He pouts.

Medium Close-Up (MCU)

MEDIUM CLOSE-UPs are the bread-and-butter shots of most on-screen conversations. A MEDIUM CLOSE-UP is close enough on the individual actor to allow her to be the full focus of attention, but it is *not* close enough to create an overly dramatic or intense mood. A MEDIUM CLOSE-UP is a focal length that frames a person from the lower chest up.

Like a MEDIUM SHOT, it is sometimes referred to as a SINGLE or ANGLE ON.

> MEDIUM CLOSE-UP—JANE. She turns to Matt and speaks.
>
> SINGLE—ANNE. She smiles in recognition.

Close-Up (CU)

A CLOSE-UP places more attention and thus more importance on the actor. A CLOSE-UP fills the screen with that person's face. It is framed from about the shoulders up and is often used to reveal important expressions or facial inflections.

> CLOSE-UP—MARK. He is shocked to hear Jane won't be coming.
> He stares for a moment, wondering whether Bill is behind this.

Extreme Close-Up (ECU)/Insert

This is as close as you can get on a person or thing without asking for a MACRO (microscopic) focal length. EXTREME CLOSE-UPs of people are usually used for dramatic impact. An EXTREME CLOSE-UP of an *item* might be used (and renamed) as an INSERT of some key item into a piece of action.

> EXTREME CLOSE-UP—JILL'S MOUTH. Her chin begins to quiver.
> She is about to cry. CAMERA TILTS UP to her eyes. The tears
> come.
>
> INSERT—The pencil on John's desk. The tip has been broken off.
> She could not have written the memo.

Two Shot

This is a shot of two people, usually in a conversation. It can be a CLOSE TWO SHOT, in which the people would probably be seen from the lower shoulders up, or a LOOSE TWO SHOT, in which most of their bodies would be on screen.

> LOOSE TWO SHOT—BILL AND DALE. They both look at the
> company manual held between them.
>
> CLOSE TWO SHOT—ANNE AND JENNY. They are nose to nose
> and furious with each other.

Over the Shoulder (OTS)

Similar to a TWO SHOT, an OVER THE SHOULDER favors one person. That means more of one person's face is showing, and most of the back or the back of the head of another person. OVER-THE-SHOULDER shots are often used in conversations in much the same way as MEDIUM CLOSE-UPS. They tend to give more of a sense of depth and relationship to the shot, however, because both parties are seen.

> OVER THE SHOULDER—JAN. She and Bill talk angrily.

Point of View (POV)

This is the point of view of some character *without* the character in the shot. As an example

> TWO SHOT—Jan and Dale. They continue to discuss the new sales policy. Suddenly the door opens. Dale turns toward it.
>
> DALE'S POV—The door swings open, and Dale's boss, BART MILLER, peeks in.

If Dale were included in the shot of the door, the shot would then be an OVER THE SHOULDER or a WIDE SHOT INCLUDING THE DOOR.

Reverse Angle

This is a shot 180 degrees from the last one. It is not necessarily a POV shot because the character may be included in the frame.

> MEDIUM CLOSE-UP—AL. He stands at the window looking outside. Suddenly, something catches his eye.
>
> REVERSE ANGLE—OVER BILL'S SHOULDER. Through the window, we see a truck pulling up at the curb.

Rack Focus

This calls for a change focus, usually radically and quickly. A RACK FOCUS typically changes the viewer's depth perspective and often juxtaposes visual elements in the frame. For instance, we may be on a shot in which we are looking at the framework of a fire escape. In the distant background, we see something out of focus, moving through the frame. We RACK FOCUS bringing the thing—two employees walking down the alley—into focus with the now *out-of-focus* framework of the fire escape framing them in the foreground. This would be written

> With CAMERA looking through the rusted framework of an old fire escape, RACK FOCUS, REVEALING two employees walking across the alley in the distance.

FG/BG

These are abbreviations for FOREGROUND or BACKGROUND. They are used in this way.

John steps past Lynn into the FG.

Ellen moves behind the bush into the BG.

Low Angle

This is an angle shot from a low position with the camera looking upward. A LOW ANGLE is often used for a dramatic, overpowering sense or to convey a feeling of dominance.

LOW ANGLE—THE BOSS. He steps into the frame holding the faulty reports and looks down into the CAMERA.

High Angle

Obviously, this is the reverse of a low angle. A HIGH ANGLE can sometimes give a sense of eavesdropping or an omnipotent perspective. It can also make the characters in the scene appear to be enclosed or trapped.

HIGH ANGLE—John. He paces in his jail cell.

Pan/Tilt

These terms call for the camera to rotate left (PAN LEFT) or right (PAN RIGHT) on its head or to tilt from looking down to looking up (TILT UP) or the opposite (TILT DOWN).

John moves to the door. CAMERA PANS with him.

John climbs the stairs. TILT UP TO FOLLOW.

Dolly/Truck

DOLLYS and TRUCKS note a forward or backward movement of the camera (DOLLY IN or DOLLY OUT) or a horizontal left or right movement (TRUCK LEFT or TRUCK RIGHT). In both cases, these moves are carried out by pushing or pulling the camera after it is mounted on a dolly with wheels.

Alice moves down the hallway. DOLLY IN past her to John seated at the table.

TRUCK LEFT following the customer into the warehouse.

Scene Heading Terms

These terms set the stage for your scene description.

Int/Ext/Day/Night

INT and EXT are abbreviations for INTERIOR and EXTERIOR; DAY and NIGHT are self-explanatory. They are the *first entries* in a scene description because they are very important to the production crew. A film or videotape production is scheduled to be shot *not* in chronological order, but rather by logistical and other considerations. Two primary scheduling considerations are times and locations. Scenes that take place at night will require special considerations and will probably be scheduled together. Scenes that happen at the same locations will also be shot at the same time, although they may appear at different places in the script.

INT JOHN'S HOUSE—DAY

EXT GAS STATION—NIGHT

EXT PLANT YARD—MIDNIGHT

Sound Terms

Sound Effects (SFX)

This notation tells all who read or work on the script that certain SOUND EFFECTS belong at specific points. For instance:

EXT BACKYARD—DAY
John tests a line with the Olson meter.

CLOSE-UP—METER FACE—The needle on the meter suddenly jumps. SFX—HIGH-PITCHED TONE generated by the meter.

Music Up/Under/In/Out/Sting

Music terms tell the editor when to bring in music to emphasize scenes or transitions.

John turns quickly toward the door. Bill steps in looking angry.
MUSIC STING.
                                                    DISSOLVE
Bill runs across the field. INTENSE, POUNDING MUSIC, UP SUDDENLY.

Sound

The word SOUND or any type of sound written into the scene description is also capitalized. This, again, is to call attention to the fact that a specific sound effect will be required.

> INT OFFICE—DAY
> John is startled by the SOUND of the RINGING TELEPHONE.

Master Scenes Versus Detailed Descriptions

As mentioned earlier, much of the technical terminology we've been discussing, especially camera terms, should be used sparingly in your script. They should be used only when you feel they are absolutely necessary to communicate your visualization. The majority of times, especially in role-play scenes, the scriptwriter should be able to accomplish this communication with the use of *master scenes*.

Master scenes are descriptions of the action in narrative terms rather than in terms of a camera's perspective. Master scenes have more of a sense of story and flow. They describe the scene with simplicity, clarity, and a sense of the proper tone, pace, and mood, but they leave the choices of angles and focal lengths to the people who are experts at selecting them: the director and the camera operator.

As you are writing your script, then, although it may be a temptation to write something like this,

> EXT FIELD—DAY
> CAMERA, looking through tree branches, REVEALS Bill in a LONG SHOT. Suddenly we RACK FOCUS to the FG REVEALING Jennifer watching John from behind the tree. MUSIC STING. CAMERA DOLLIES around to a CU of Jennifer, as she stares at Tom with anger.
>
> REVERSE ANGLE—OVER JENNIFER'S SHOULDER. We see Bill turn and discover Jennifer watching him. He moves toward her.
>
> CLOSE-UP—Bill. He glares back and finally speaks.
>
> > BILL
> > Well. Are you happy now?
>
> MEDIUM CLOSE-UP—Jennifer
>
> > JENNIFER
> > I've been unhappy for months now and
> > you know it.
>
> MEDIUM CLOSE-UP—Bill. DOLLY IN to EXTREME CLOSE-UP.
>
> > BILL
> > So have I.

you should write this instead.

EXT FIELD—DAY
Jennifer hides behind a tree watching John move along the path
in the distance. Her eyes hold on him with controlled anger.
Suddenly, Bill turns and sees Jennifer watching him. He is
furious. He walks up to her and looks directly into her eyes.

> BILL
> Well. Are you happy now?

> JENNIFER
> I've been unhappy for months now and
> you know it.

> BILL
> So have I.

The first version of this scene not only is cumbersome and difficult
to follow but also places the director in a visual box that he will probably
not care for.

The second version is a master scene. It is clear, simple, and descriptive. Better yet, it allows the director—the person hired to commit the
script to tape or film—to make choices about the camera positions and
visual aesthetics. She would probably make these choices based on a
number of factors, including the emotional content of the scene as the
actors worked with it on location.

Then again, you may be writing a script in which detailed camera
directions are *helpful* to the director.

INT SHOP—DAY
Back with the host at his
workbench. He removes the cover
from the E-17 meter and tilts it
forward, propping it up on a
wedge so that we can see the dials
and VU meters.

> HOST
> As you can see, the E-seventeen is
> a compact unit with number dials,
> meters and switches on its
> face. . . .

As the host points out:

> HOST

ECU—on/off button
The on-off button is located here
in the upper right.

ECU—Polarity switch
Beside it is the polarity switch.

ECU—VU meters
In the center are two VU meters.
VU stands for volume units. As
you can see, they indicate

PAN VU-METER SCALE
readings of minus one hundred to
plus ten.

In this situation, the suggested ECUs or EXTREME CLOSE-UPs are appropriate. They call the director's attention to the proper areas of the meter that should be seen closely during the corresponding narration. In the case of the VU-meter reading, they indicate that a PAN across the scale would work well to visualize the numeric markings. Whether or not the director actually uses a PAN here is up to him. He is made aware, however, that seeing the numeric scale is an important part of the visual content of the program, and a PAN is one effective way to accomplish this.

Another situation in which the use of detailed scene descriptions might help both the producer and director would be in the development of a script that calls for a distinct visual tone, pace, or look.

The following opening is from such a script.

NOTE: PORTIONS OF THIS COMMERCIAL SHOT IN THE GYM ARE RECORDED WITH A CONSTANTLY MOVING, HAND-HELD CAMERA. THE PHOTOGRAPHY IS BLACK AND WHITE, HIGH-GRAIN, AND HARSHLY LIT. THE IMAGES ARE SEEN IN JERKY, SAMPLE FRAME MOTION. NONGYM SCENES ARE RECORDED NORMALLY.

                                                          CUT IN
INT GYM—DAY
As MUSIC BANGS IN, a sweaty female bicep FILLS THE FRAME. It curls upward, straining.

CAMERA FOLLOWS worn, gloved, female hands as they slam a 25-pound weight onto the end of a barbell.

OVERHEAD ANGLE—The top of two glistening, flexing female thighs. The legs extend pushing against Nautilus-type foot pedals.

CLOSE-UP—Two female feet hammer rapidly against the floor in an aerobic work out. We TILT UP QUICKLY and see the exhausted face.

MUSIC CUTS OUT ON . . .
                                                  DVE STARBURST TO
INT OFFICE—DAY
The same female we've been watching ''torture'' herself in the gym is now seen calmly working in her office. She is in business dress and looks slender, healthy, and attractive—not in any way tough or bulky. She looks up from her work and speaks to an OFF-CAMERA INTERVIEWER:

                    FEMALE EMPLOYEE
                     (Matter-of-factly)
            All I can say is I get something there I
            can't get anywhere else. It's incredible.

As is obvious, in this instance the script is suggesting a specific look. Even in this case the director may choose to record the scenes differently, but at least he has been given a precise visual "feel" for the tone and pace the client and producer obviously want.

Before attempting to develop scripts like this, you should be *absolutely sure* you are familiar with the technical and aesthetic aspects of what you are suggesting.

The Writer-Director Relationship

In the end, the amount and types of technical terminology you include in your script come down to that same elusive element we mentioned earlier, your intuitive sense of what works in each case.

This sense, as we have said, comes from experience. In this case, part of that experience includes your awareness of the writer-director relationship, which states that the writer's job is to provide the narration, dialogue, and suggested visualization. The director's job is to make the writer's script work on the screen. If, based on her experience, she feels the visualization should be different than what you've described, she has the authority and in fact the *obligation* to change it. This also includes the dialogue or narration.

Why, you may wonder, should a director be able to change what you have spent weeks researching, a script on which you've agonized over every word, a script your client and producer have both previously approved?

The answer is simple and perfectly logical. When things are written on the page, regardless of the time and effort that went into their conception, they may not work the same way on the set or on a location. An actor may not be able to say certain words or lines, a particular angle may not cut properly in editing, or a sound effect may not be obtainable within the budget or time constraints of the shoot. As the person ultimately responsible to be sure that all elements of the production process have been carried out efficiently and effectively, the director becomes the logical person to make final decisions in these instances.

Although some new writers are not totally comfortable with this relationship, you should abide by it. As you become more and more familiar with the production process, you will come to understand why it has evolved this way. Until then, write with great clarity but in a way that allows the production specialists the maximum flexibility in actually recording your vision.

■ ■ ■ ■ ■ ■
You Script It 7

A corporation has decided it needs a program to communicate the new rules of business survival. Things, it seems, are getting increasingly tough and competitive in the marketplace, but employees aren't getting the message. According to the managers you are now speaking with, there is a relaxed, unhurried attitude that pervades the lower levels of the organization when what's really needed is a tough sense of urgency.

They have called you in to develop a program concept, and you are halfway through your initial meeting. Your design information so far indicates the audience is mostly male, high-school-educated, blue-collar workers. The age range is roughly 25 to 45, and there is a secondary audience of first-line supervisors.

As you listen to the meeting participants discuss the company's predicament, the words that keep coming up over and over are *tough* and *competition* and *survival*.

You are still in the midst of extracting your design information when one executive suddenly looks at his watch and says, "Look, I realize you'd probably planned to go home and think about all this, but I've got a meeting in 15 minutes with my boss, and it would really help if I could bring him something, anything! Any basic ideas or concepts?"

What would you say?

Dialogue 10

□ □ □ □ □

With a good grasp of basic script formats and the common terminology they incorporate, you now need to consider two other major script elements: *dialogue* and *narration*.

Just as your visualization is critical to the *picture* part of your presentation, dialogue and narration are the critical *sound* part.

Critical, by the way, is not an overstatement. If your visuals are brilliantly thought out but your dialogue or narration is weak, the script will not work. In a well-written script, the audio supports the visuals and vice versa. In most scripts, the two elements should be interwoven like the tightly knit strands of a creative fabric.

The trick is to write visually stimulating scripts that add another dimension to the words.

—Gary Schlosser
Corporate Executive Producer

Words and Pictures

I use the words *most* and *should be* because, like the script format itself, the relationship between sound and pictures and the relative importance of either one in a script are *not* constant. The importance of words and pictures varies with the type of script you may be writing.

Consider two examples: an executive's narration script and a script for a motivational, historical video.

In the first case, although it is important that we see the executive on camera, when developing his script the visual element means little or nothing to the writer. The tape will simply be a talking head in a three-piece suit, most likely seated in a high-backed office chair. As a writer, you would first strive for clear, powerful narrative in this script and give secondary thought to the visualization.

In the second case, the pictures would be of prime importance. Granted, words and music are also very important in a historic presentation, but this would primarily be a visual program focusing on what the past looked like. In this case, you would probably visualize each scene first and write in narration as a secondary function supporting those visuals.

With this word/picture relationship in mind, then, let's first establish a definition for *dialogue: the words spoken in a conversation between at least two characters in a role-play situation.*

Well-written dialogue adds credibility and content to your script. In other words, it is believable, and it gets the message across. Let's look at a few examples of how these two factors result in effective dialogue passages.

Case Study 10   *One Picture Can Say More Than. . . .*

One writer I know found a memorable way to close a talking head executive program. The purpose of the program was to convince employees they needed to develop an increased sense of urgency because competition was eating away at the company's market share.

The program was to be created primarily from videotaped speeches a group of executives had given in a recent seminar for key managers.

In looking through footage of the speeches as part of his research, the writer found an executive who said, "And what happens if we don't get busy right away?" The writer chose to make this the last line of the program and called for the executive's head shot to dissolve to

> EXT PARKER STREET YARD—DAY
> This is the yard as seen 5 years in the future. A rusty chain-link fence has been put up blocking the entrance. It is padlocked shut. Beyond it, the yard itself is ghostly, devoid of trucks, equipment, or people. What it does have are lots of weeds, paper trash, and clumps of grass growing up through the cracked tar. SFX: A SLIGHT WIND CAN BE HEARD, NOTHING ELSE.
>
> CAMERA HOLDS, THEN PANS. Just to one side of the gate, we see a real estate sign on a post, which has been hammered into the dry ground. It offers the company yard for sale as a "Commercial Bargain."
>
> AS CAMERA PUSHES IN SLOWLY, we . . .
>
> FADE OUT

The producer liked the idea so much that she arranged to shoot it on a weekend. With the help of the fleet group, she managed to move all vehicles in the equipment yard out onto a side street. Tumbleweeds, paper trash, and fake grass clumps (actually small plastic ferns) were brought in, as well as the ominous real estate sign.

When the program was completed, many who saw it felt that the single visual image of the abandoned yard "said" more in a few seconds than all 35 minutes of executive speeches put together.

Dialogue Credibility

Assume we are writing an instructional, role-play script on telephone sales. In one scene we are working on a piece of dialogue between a new, somewhat timid female sales rep and a friendly, experienced male rep. The female rep is frustrated because she can't seem to get the hang of the work. She decides to confide in the male rep.

In order to provide credibility, the dialogue should simply sound natural, as if it could really have been spoken between these two characters.

One way of handling this conversation might be scripted this way.


```
FADE IN
INT OFFICE—DAY
JIM, a friendly, experienced
telephone sales rep is having a
cup of coffee with JAN, a timid
rep now in her third day on the
job.
```

                    JIM
          So how are you getting along,
          Jan? Are your telephone sales
          skills effective yet? Sometimes
          with new people the skills need a
          little time to develop. I think if
          you are having trouble, a little
          experience will help you a great
          deal.

                    JAN
          No, I am afraid they are not. I
          just cannot seem to make a sale. I
          am not sure I know what I am
          doing wrong, but I think I am not
          showing enough initiative.

                    JIM
          Try the three sales tips we all use
          here. One is pitch the sale. Two is
          overcome objections, and three is
          close the sale. I hope these help.

                    JAN
          Thank you. They will.

A second version of the same interaction might be scripted this way.


```
FADE IN
INT OFFICE—DAY
JIM, a friendly, experienced
telephone sales rep is having a
cup of coffee with JAN, a timid
rep now in her third day on the
job.
```

                    JIM
          So what's up, Jan?

> JAN
> (shakes her head)
> Problems.

> JIM
> Trouble making sales?

> JAN
> More like a sales disaster!

> JIM
> (chuckles)
> Well, don't get rattled. Sometimes
> it takes a while to get up to speed.

> JAN
> (depressed)
> I don't know, Jim. I think I'm
> just not assertive enough.

> JIM
> Have you tried the three ''golden
> rules'' of telephone sales?

> JAN
> ''Golden rules?''

> JIM
> Yeah. Sounds corny, but they
> work.

> JAN
> I'll try anything at this point.

> JIM
> Well they're basically what I call
> POC—pitch the sale, overcome
> objections, and close the sale.

> JAN
> (thinks)
> Hm . . . POC. Pitch, overcome, and
> close.

> JIM
> Try 'em out. They'll help.

> JAN
> Okay.

I think you'll agree that the second version is much more credible. But why? What specifically are the qualities that make one version sound very stilted and unnatural and the other sound like a real conversation? There are actually three.

> natural speech patterns
> believable character motivation
> use of contractions and colloquialisms

Natural Speech Patterns

Natural speech patterns mean that the conversation is kept moving back and forth between the two characters, often after only a sentence or two. At times, the sentences themselves are actually sentence fragments. This is the way people naturally express themselves.

In the first version, notice that the speech patterns are very *unnatural*. Jim's opening lines contain four separate thoughts clustered together. Rather than *converse* with Jan and allow these ideas to emerge naturally, he simply blurts them all out at once. Jan returns this unnatural delivery with her own thought cluster. The result is a very awkward exchange that actors would find difficult to say convincingly and viewers would find even more difficult to believe.

Believable Character Motivation

Volumes have been written on the intricacies of character development and motivation alone. You will explore these intricacies over the years at deeper and increasingly complex levels as your writing skills broaden and mature. As a new scriptwriter, begin with the simple ideas that character is the basic *personality* of an individual and motivation means *purpose*, a legitimate reason for doing something. This purpose can come to a character from any number of sources. It might be a phone call, a question, a hunch, or a headache. It could be an emotion, such as fear or love, a question about the future or a memory from the past, or just the type of personality you've decided she should have. The main things are that it be believable in the situation based on the character's personality and that it help move the story forward.

In the second version of our scene, both these items are accomplished. The story is certainly moved forward by the dialogue and the friendship that is struck up as a result of it, and the characters seem motivated to speak as they do because of the *situation* and *personality traits* we've given them. I imagine Jan as somewhat timid, depressed, and probably wanting to do a good job, but feeling frustrated because she can't. I find it perfectly believable that a woman like her would probably be *motivated* to tell her troubles to a sales rep with a lot more experience who seems to have a sense of concern for her predicament.

Jim appears to be a friendly and open type. I can imagine him being very good on the phone as a sales rep. He also seems compassionate enough to be *motivated* to lend a hand to a new person in need of not only the "golden rules," as he calls them, but also a little plain old support.

In the first version, the characters seem lifeless and robotic. They project almost no characterization. I can imagine them facing one another, and simply letting their mouths mimic the words being spoken with no warmth or sense of human emotion at all.

Try to tell your story in terms of people. It's easier to get the audience to care about people's lives than about hardware or statistics.

—Alan C. Ross
Corporate Writer/Director

Use of Contractions and Colloquialisms

The first version of this scene contains no contractions. This is another unnatural way of speaking. In casual conversation, most people usually don't say "I am," "do not," or "cannot." They say "I'm," "don't," and "can't."

In the second version, both characters use contractions in ways that have human qualities. They also use typical casual and colloquial phrases like "Try 'em out" and "up to speed."

What all these elements add up to are natural human speech qualities that ring true (credible) to our ears.

Dialogue Content

Besides this element of credibility, in most corporate programs effective dialogue must also provide content. It must make a point. If it doesn't, no matter how great it may sound, it boils down to wasted screen time at a very expensive price.

In effective dialogue, the content is worked into the conversation naturally. To see how this is done, let's look at another example. In this case, the subject is quality. The scene involves two assembly line workers and the critical content points are:

1. Quality is achieved by following strict standards.
2. Quality can be measured by using those standards as "yardsticks."

The first version is scripted this way.

```
FADE IN
JIM and JAN, two assembly-line
workers, are at lunch.

                                        JAN
                        Jim, how's it going?

                                        JIM
                             (cynically)
                        Great! I just got nailed because
                        my "quality" is falling off.

                                        JAN
                        Your what?

                                        JIM
                        My "quality."

                                        JAN
                        I don't get it. You punch gear
                        holes on a metal stand. How does
                        that relate to quality?

                                        JIM
                        Bill says quality is standards. I've
                        been working for this place for
```

three years now, and nobody's
ever bothered me about that.

                    JAN
What standards?

                    JIM
The yardsticks, according to Bill.
You been here for longer than me.
You ever been approached about
how many parts are rejected with
your number on 'em?

                    JAN
Nope. Guess I better start
watching out, though.

A second version might read this way.

FADE IN
JIM and JAN, two assembly-line
workers, are just leaving their
areas for lunch. They approach
each other, walking toward the
parts reject bin.

                              JAN
                    Jim, how's it going?

                              JIM
                         (cynically)
                    Great! I just got nailed because
                    my "quality" is falling off.

                              JAN
                    Your what?

                              JIM
                    My "quality."

                              JAN
                    I don't get it. You punch gear
                    holes on a metal stand. How does
                    that relate to quality?

                              JIM
                         (still very cynical)
                    According to Bill, in two ways.
                    One is that he says the only way
                    we can get quality is by following
                    standards.

                              JAN
                         (A hint of anger)
                    Such as?

They are just now passing the
rejects bin. Beside it is a
''standards'' sheet hung on the
wall. The sheet lists proper
tolerances and ratings figures for
the parts they work on. Jim
motions for Jan to stop. He then
points to the sheet.

                    JIM
          These, the measurement
          standards: dimensions, tolerances,
          polish grades. That stuff.

                    JAN
          So if your parts are meeting these
          figures, they're okay?

Jan looks at the sheet.

                    JIM
          Right. Up to standards. And that
          means they're good quality.

                    JAN
                  (sly smile)
          And just how does good old Bill
          know how your parts are
          measuring up?

                    JIM
          That was the second part of our
          little, ah, ''discussion.'' He uses
          the standards as a yardstick.

                    JAN
          Yardstick?

                    JIM
          Yeah. He checks out what
          percentage of my parts are
          meeting those standards and uses
          that to measure my performance.

                    JAN
          You mean they really do measure
          your performance rating against
          the standards?

                    JIM
          Right. But not just me. All of us.
          You gotten any flak yet?

Jan picks up a part out of the
rejects bin. She looks it over and
then drops it back in the bin. The
two start off together.

                    JAN
          Nope. Guess I better start
          watching out, though.

The first version of this exchange sounds fairly natural because it incorporates the basic speech elements just discussed. It does not, however, get the point across. Although the two content points are present, they lack focus and motivation. They seem like unnatural bits of information just dropped into the dialogue. The focus, instead, is on how long Jim has worked at the company and how he's never been bothered about quality before.

In addition, the scene takes place at lunch. This is a natural enough setting for such a conversation to unfold, but in this case it's *not* the best place to use support visuals as motivation for natural, content-effective dialogue.

The second version solves these problems in several ways. First, it focuses on the proper content points as a primary reason for the conversation. Because of this focus, the content points become a natural and motivated part of the dialogue. To help accomplish this, the emphasis on Jim's tenure with the company has been dropped and both employees are characterized as more cynical about the company's scrutiny.

Second, the visuals provide solid content support. This is accomplished by having the scene take place while the two are passing by the rejects bin. The standards sheet is close by on the wall to act as a source of motivation for Jim to explain just what standards mean. It also prompts Jan to ask, "You mean they really do measure your performance rating against the standards?" The rejected part in the bin provides more natural motivation. It prompts Jan's last line about starting to watch out.

The Dialogue Test

Like the many other aspects of scriptwriting, good dialogue skills come with practice. One of the best ways to polish your dialogue skills is to listen closely to the way people speak. Also, speak the dialogue you write out loud. Never write mindless dialogue, though. Always write it to communicate some point or character aspect. This will help you achieve focus as an inherent part of your writing habits.

As you go about this process, there are some points that may help to guide you. A test of good dialogue would contain positive answers to the following questions. Until you become very comfortable and accomplished at writing dialogue, you should ask yourself each question about every exchange you write.

1. When spoken out loud, does it sound natural?
2. Does it emphasize the proper content points in a way that will be believable to the viewer?
3. Does it prompt the viewer to feel that the actions and personality of each character are credible and motivated?
4. Does it incorporate visual elements in the script to help convey as much content and characterization as possible?
5. Is it simple enough to convey the content clearly?

If your dialogue continually passes this test, it will be a valuable addition to your script. It should also please your producer, director, client, and, of course, the actors whose job it is to present your words believably on the screen.

Dialogue, however, is not the only verbal way to convey information. As we've previously mentioned, another method is with the use of narration.

■ ■ ■ ■ ■

You Script It 8

If a producer asked you to clean up the poorly written dialogue in the following scene but keep the content unchanged, what two things would you do to it?

Write your new dialogue, and think about specifically *what* you've changed and *why*.

```
INT OFFICE—DAY
Jeff and Mary continue their conversation.

                    JEFF
          I have seen you handle the paperwork.
          You do the mathematics as a first step,
          then the envelope typing. I am excited
          about doing it myself. I promise to do a
          good job.

                    MARY
          It is good that you have watched me do
          the forms. I am glad you are excited,
          and I am not worried at all. You will do
          just fine.
```

Narration 11

☐ ☐ ☐ ☐ ☐

As we have seen, much of the information your audience will walk away with will be communicated by words. In the case of narration, those words are *not* spoken by one character to another in a role-play situation. Instead, *narration* could be defined as: *words spoken by a voice-over narrator or on-camera spokesperson, directly to the audience.*

Effective Narration Qualities

To be effective, narration should also provide content and credibility. Effective narration usually accomplishes this with three general qualities.

conversational tone
content focus
simplicity

Conversational Tone

Conversational means *informal.* Conversational narration should sound as if it were being delivered off-the-cuff rather than as a prepared script.

The importance of this impromptu tone becomes clear when you consider that a narrator—especially an on-camera narrator—must establish a rapport with the audience if she is to be credible. The audience must like and identify with her in order to accept the material being presented. If what she says sounds prepared or *un*natural, she immediately becomes suspect. Once this happens, achieving that audience rapport will become difficult at best.

What can you do as a writer to build this conversational tone into your scripts? The most important rule is simply to write to be heard, not to be read.

Again, let's illustrate by considering two versions of the same script. In this case, the subject is how to become a successful purchasing agent. The first version is scripted this way.

```
FADE IN
A HOST steps into the FRAME in
black limbo. He turns to CAMERA
and says . . .

                              HOST
              Purchasing is a difficult business
              in which to achieve success.
              Regardless of this difficulty,
```

however, being a successful
purchasing agent is indeed
achievable if the proper
preparatory steps are taken. This
videotape program will present
those steps. They are: 1. Know the
marketplace. 2. Shop for prices.
3. Bargain for deals. 4. Obtain the
prices in contractual form.

A different version of the same material might be scripted like this.

FADE IN
A HOST steps into the FRAME in
black limbo. He turns to CAMERA
and says . . .

          HOST
It's tough to become a successful
purchasing agent. There's lots to
learn, and it's a competitive
business. But you <u>can</u> be good at
it if you follow the right steps.
Care to learn how? If so, listen
up, because today I'd like to talk
with you about those steps and
hopefully get you started down
that road to success. We'll be
looking closely at: knowing the
marketplace, shopping for prices,
bargaining for deals, and getting
those deals written in a contract.

The first version of this piece is definitely formal and written more
to be read than to be heard. It uses many polysyllabic, rhetorical sounding
words like *difficult, achievable, indeed, proper, preparatory,* and *however.* It also uses no contractions, and it lists out the steps to be followed
exactly as they might have been listed in some *manual.* In fact, this
entire piece sounds like it might well have been lifted straight from the
pages of a manual.

Because of these qualities, the host reading this script will have a
difficult time making it work. No doubt the film or taping session will be
stopped frequently for discussions between the actor, client, and director
in an effort to make the material "sound better." If it does get recorded as
is, the audience will probably lose interest in it quickly and retain very
little of what's been presented.

The second version, by contrast, has a warm, human tone. It uses
much simpler, commonplace words and colloquial phrases like "lots to
learn," "start down the road," and "be good at it." It also uses contractions

frequently, and it addresses the audience directly with an occasional reference to "you."

The second version uses another natural speech quality we've already discussed: the use of questions. Besides sounding very "human," an occasional question helps draw the audience into the program.

The result of all these qualities is narration an audience will listen to and learn from, conversational narration written to be spoken and heard, rather than read.

Content Focus

We touched on the importance of focus in our chapter on dialogue. The same rule holds true for narration. It must not only sound good but also make the point. In order to accomplish this, you must focus on the proper content points at the proper times. For the most part, this means being sure the points are present in the material, they are clearly stated, and the focus is on them versus other aspects of your script.

As an example, in the second version of the script passage on becoming a purchasing agent, the host makes a point that it's "tough to become a successful purchasing agent. There's lots to learn, and it's a competitive business." With this statement, an important point is made very briefly. In addition, the audience is set up for the subject about to be discussed. The narration then moves on to the subject—learning the proper steps—that is the "hard" content for this bit of narration.

Were the narration to have continued to *focus* on the subject of how tough it is to succeed in the purchasing agent business, however, whether it sounded conversational or not, it would have lost its value to the audience. Imagine it reading like this.

> It's tough to be a purchasing agent. There's lots to learn, and it's a competitive business. Besides that, the hours are long, and the working conditions aren't the greatest. Match that with the poor pay and . . . . Well, you get the picture. But you *can* be good at it, if you follow the proper steps.

As you can see, this sounds conversational enough, but . . . well, you get the picture.

Simplicity

The need for clear, simple English can't be overemphasized in any form of writing. In script development, that need is even more critical.

In a film or videotape, words go by quickly. Granted, they are supported by pictures, but they are also heard for only an instant and then gone. Furthermore, unlike the reader of a book, a program viewer can't simply go back and reread that section as she might a written passage.

You should, therefore, strive for clear, simple word usage in every script you write. Although this applies to narration, it also includes your dialogue and scene descriptions.

Simplicity means reducing a subject to its lowest descriptive common denominator. It also means letting visuals support the spoken words. To illustrate, let's consider another example.

The subject, in this case, is the path a service order form takes from the time it is written by the customer representative until it's completed and filed. Here are two ways that process might be scripted.

INT SET
We now find the hostess seated at a customer representative's desk. She looks up from a stack of service orders. Holding one in her hand, she gestures and says . . .

> HOST
> Now that we've established what a service order is, let's look at exactly how it moves through our system.

INT SERVICE OFFICE—DAY
A customer rep on the phone with a customer. She is just finishing filling out an order.

> HOST (VO)
> First, it's filled out by the customer rep. At this point it's in four copies.

The customer rep hangs up and tears out the carbons, separating the order into four parts.

> HOST (VO)
> Each copy plays an important part in making sure the order is completed efficiently and on time for the customer.

INT WAREHOUSE—DAY
A packing and shipping clerk receives his copy of the order and heads into the warehouse to locate the material. His copy is white.
SUPER: WHITE/SHIPPING

> HOST (VO)
> The first copy—the white—is sent to shipping. Here, the clerk uses it to locate the material and

EXT WAREHOUSE PARKING LOT—DAY
The same clerk loads an equipment box on a truck.

> HOST (VO)
> . . . load it on the van routed for that part of the city.

INT BILLING OFFICE—DAY
A billing clerk uses her copy—the pink—to type billing information into a computer.
SUPER: PINK/BILLING

> HOST (VO)
> The second copy—the pink—is sent to billing. Here a representative enters the charges into our system.

EXT A HOUSING TRACT—DAY
The delivery clerk is on the road
later in a tract. He uses his copy
of the order to find the customer's
address. He stops at the right
house and heads for the front
door.
SUPER: YELLOW/DELIVERY

HOST (VO)
The third copy is the yellow one.
It's used by the delivery clerk to
locate the customer's house and
finish the delivery.

INT SERVICE OFFICE—DAY
Back to the original customer rep.
She has routed the other copies
and is filing the last—blue—copy
in a large file cabinet.
SUPER: BLUE/FILE

HOST (VO)
And finally, the blue copy is filed
by the original rep who took the
order. It's kept for three years.

This is a simply scripted group of scenes that clearly illustrates the
path a service order takes. The selection of what facts to include is
correct, the word usage is simple and direct, and the narration is supported by appropriate visuals, including support titles. Anyone watching
this should have no trouble following the words or the action and absorb-
ing the material.

But suppose the script went like this.

INT SET
We now find the host seated at a
customer representative's desk.

HOST
Now that we have established
what a service order is—a vehicle
to communicate the various
aspects of equipment purchase
and delivery—we must examine
its method of processing in a
timely fashion through the
various channels of our routing,
filing, and field delivery systems.

INT SERVICE OFFICE—DAY
A customer rep on the phone with
a customer.

PRODUCT SHOT—four-part form
spread out.

HOST (VO)
First, the order is filled out by the
customer rep. At this point it is a
four-section form combined into
four multicolored, carbon-
interfaced copies. The multiple-
copy system helps with speedy
distribution of the paperwork.
Each part plays a critical role in
making sure the order is written,
routed, delivered, and finally
completed efficiently and on time
for the customer.

INT HOUSE
Customer with product. She is
pleased.

INT SERVICE OFFICE—DAY
Rep files blue copy.

INT WAREHOUSE—DAY
A packing clerk receives his copy
of the order.
ECU—ORDER

PRODUCTS ON SHELVES

This critical role will continue to
assure our success in the
marketplace and solidify our role
as a leader in product
applications.
The fourth, blue copy is filed by
the customer rep and kept in the
hard-copy file system for a total of
3 years, whereas other copies are
routed throughout the company.
The first, white copy or top copy
is sent via company mail systems
to shipping. Here the clerk uses it
as a search source. Using the
detailed information provided by
the original customer rep who
wrote the order, he attempts to
locate the material on shelves
coordinated by number to the
parts and shipping list.

As you can see, this version of the same material is hardly simple, direct, and clear. On the contrary, it breaks nearly every rule of clarity in the book.

The word usage is stiff and formal, with constant use of clumsy phrases like "multicolored, carbon-interfaced copies."

This type of description throws the viewer for a loop the instant he hears it. While he is trying to figure out exactly what that mouthful means, the program moves on to other (in this case, poorly written) examples.

One is the constant lack of focus. Another is the long reference to the service order copies themselves, ending with the fact that proper service order processing will "continue to assure our success in the marketplace and solidify our role as a leader in product applications."

This idea may be fine to include in a program on competition or strategic positioning in the marketplace, but it has no place in a program on the steps involved in service order processing. It simply distracts the viewer's thought process from the real issue at hand.

Other violations of the rules of simplicity in this script include things like long, cumbersome sentences. Just one example is: "Using the detailed information provided by the original customer rep who wrote the order, he attempts to locate the material on shelves coordinated by number to the parts and shipping list."

Again, this type of writing stops the viewer as she tries to figure out exactly what it means. In the meantime, the show goes on without her.

Structure is still another problem in this script. The natural order a sequence like this should take is leapfrogged in this version by covering what happens to the *fourth* copy as we leave the *first* part of the discussion. We'll have a closer look at structure in the next chapter, but, briefly, this back-and-forth thought process causes the viewer to make even more difficult mental gear shifts. He must first orient himself to exactly where we are in the process when we leave the first part to go to the fourth. He must then reorient himself when we leave the fourth part to return to the second!

Last but not least, the visuals are weak in this version. Scene descriptions are vague, sketchy, and poorly thought out. As an example, the first scene description places the customer rep on the phone but doesn't give her any action to support the narration. She is simply talking. Right after this, we are given a shot of a service order separated into four parts. If the program plays out as scripted, this shot will cover a block of narration that runs approximately 30 seconds. Holding on four sheets of paper for this long will seem like an eternity on the screen. In addition, support titles that would reinforce the idea of the color-coded order distribution are not included.

Finally

As we've seen, numerous elements are involved in well-written dialogue and narration. Those we've covered are really only the basics. You will move naturally into more sophisticated aspects of dialogue, characterization, and narration as you gain skill and experience, but don't expect to master them overnight. Instead, you should strive for continual improvement and practice the basics every chance you get.

Think of a topic, give it three logical steps, and write the words you imagine a program host or two characters might say. Speak those words into a tape recorder and listen to them. Note which words come out naturally and which cause verbal stumbling. Also ask yourself if the words are clear, well organized, and descriptive. Are they flat and passive or crisp and active? Most of all, do they sound like the natural language of real people?

■ ■ ■ ■ ■ ■
You Script It 9

Many corporate programs can be thought of as a mixture of show (visual) and tell (nonvisual) communication. As an example, a typical program might be outlined on a show-and-tell chart this way.

Show	Tell
Opening montage. Employees on the job.	Main title appears.
	Host enters, tells viewer the subject, three safety rules.
Vignette showing first safety rule as it should be carried out.	Support titles or graphics.
	Host reappears and introduces second rule.
Vignette. Second rule is seen being carried out properly.	Support titles of graphics.

	Again, host appears. He now introduces third rule.
Vignette. Third rule carried out.	Support titles and graphics.
	Host reappears. He summarizes and wraps up the show. Support titles may also be used.
Closing montage and company logo/credits.	

As you can see, the host and support titles could be considered the "tell" (nonvisual) parts of this program. Assuming this is true, how could those two elements be made more visual and thus become at least partially "show-" type elements?

Structure and Transitions

Corporate films and videotapes are produced to train, motivate, and inform. To accomplish this, they must impart information in ways that make it easily understood. In other words, they must have what I often call a high clarity factor.

The level of that clarity factor is directly affected by all the script elements we've been discussing: concepts, visualization, dialogue, narration, format, terminology, and so on. It is also affected by two other, perhaps more subtle elements: the *structure* of the script and the *transitions* used to move the viewer through its scenes.

You might think of the relationship between these two elements in terms of an analogy: An informational or instructional script could be considered a series of content islands the viewer must visit that are linked by a string of transition bridges.

With this relationship in mind, script structure could be defined as: *the arrangement of content information into logical, easily understood script segments.* Script transitions are the sound or picture elements linking script segments.

Types of Structures

There are many types of program structures. Often they are based on story line and the information being communicated. For instance, a story about a supervisor who gets reprimanded and nearly fired might include the events that led to his predicament. This story could be told effectively in a series of flashbacks. If so, it could have a back-and-forth, hopscotch type of structure. Information could be conveyed while transporting the viewer between the present and the past. In this type of story, the transitions might be the special visual effects—perhaps slow dissolves—that mark our departures from one time frame to another.

Basic dramatic structure usually follows a series of escalating crisis situations that eventually lead to a climax and finally a resolution (Figure 12.1, page 116).

For instance, a protagonist is faced with a conflict or dilemma. When he tries to resolve it, the first crisis materializes. He overcomes this one, only to find another, perhaps even worse, barrier now in his way. Again he triumphs but is faced with still another crisis. The tension builds with each of these challenges until it eventually becomes a do-or-die situation. In the climax, the protagonist overcomes the final big one against all odds. We are then given the resolution: the antagonist's ruin, as the protagonist finds peace. Although many dramatic stories incorpo-

*Y*ou are dealing with a visual medium. When organizing your material, first try to lay out the entire story without using any words. This will point out where your visuals are weak.

—Alan C. Ross
Corporate Writer/Director

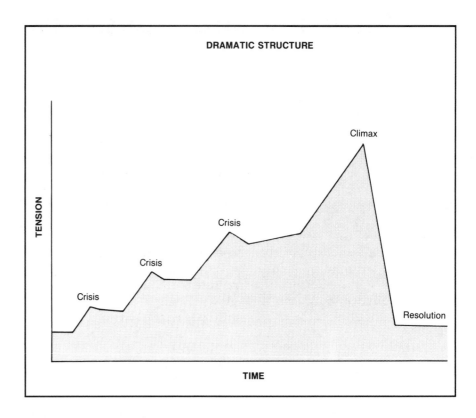

Figure 12.1 Basic dramatic structure.

rate different story lines, characters, and plot elements, most follow a dramatic structure similar to this.

Transitions in such a structure are special effects, sound and music cues, dialogue, and combinations of these. For instance, in the story just mentioned, at certain crucial points the music might build to dramatic bridges as we dissolve to upcoming scenes. This music device would act as a kind of subconscious signal to us that a dramatic change is about to occur.

Structure Versus Story Line

Structure should not be confused with concept or story line. One structure might work better in a certain kind of story line, but the structure is simply the arrangement of the information, not the story "formula."

For instance, our story about the nearly fired supervisor might seem like a natural for a flashback structure, but it could also be told chronologically. In this case the basic program concept or story formula—supervisor goes bad, gets caught, and is reprimanded—remains the same in both. The arrangement of how the story information is communicated, however, is changed.

Corporate Structures

Corporate programs can utilize any script structure that is effective. The type of structure you choose for a project should be heavily influenced by the amount and type of information to be conveyed. Because information is usually of prime importance in corporate programs, some follow the simple, time-proven structure briefly mentioned earlier in which facts are easily communicated:

Tell 'em you're gonna tell 'em.
Tell 'em.
Tell 'em you told 'em.

This type of script structure is a very effective way to present technical material for training or informational purposes. It is by no means the only way to convey information in corporate programs, but you should be familiar with it, especially for writing technical or instructional scripts.

Tell 'em Examples

We can best illustrate both the use of this structure and the basic value of good structure in general with another script example. In this case, we'll assume the content involves communicating a technical overview of a three-step process used to repair electronic tool-cleaning booths, known as Electrostats. The approach is a traditional voice-over narrator with example footage. Here's the first version.

                                                          FADE IN
INT CLEANING BOOTH—DAY
A REPAIR TECHNICIAN enters the booth, removes his equipment,
and begins to perform tests on the Electrostat.

> NARRATOR (VO)
> The first step is to test the two circuits:
> the secondary circuit should be tested
> after the primary circuit, using all test
> methods that are in the first step.

INT CUSTOMER'S OFFICE—DAY
The customer receives a phone call. She looks pleased.
SUPER TITLES: FAX OR MAIL ALSO

> NARRATOR (VO)
> After completing the two steps, the
> customer should be informed by mail,
> telephone, or fax. This is the last step
> or step three.

INT CLEANING BOOTH—DAY
The repair tech is soldering a new coil into the circuit.

> NARRATOR (VO)
> The second step is to repair the circuit.
> The secondary circuit or the primary
> circuit can be repaired on the spot or

INT CLEANING BOOTH—DAY
A different repair tech places a referral notice instead of fixing
the station.

>   NARRATOR (VO)
>   . . . by adding a second function to this
>   step, which is a referral step on
>   another date.

INT OFFICE—DAY
The repair tech is calling the customer seen earlier on the phone.

>   NARRATOR (VO)
>   Step three should take place after
>   repair. It involves informing the
>   customer, as mentioned in step two.

INT BOOTH—DAY
ECU of the referral notice placed on the equipment.

>   NARRATOR (VO)
>   This, however, should not take place if
>   the second part of step two—referral to
>   another date—is used.

Now here's our second version of that same procedure.

>   FADE IN
INT CLEANING BOOTH—DAY
A REPAIR TECHNICIAN enters the booth, removes his equipment,
and begins to perform tests on the Electrostat.
SUPER TITLES: TEST, REPAIR, INFORM

>   NARRATOR (VO)
>   Testing an Electrostat workstation
>   involves three steps: testing, repairing,
>   and informing the customer. Let's start
>   at the beginning.
>   DISSOLVE
INT CLEANING BOOTH—DAY
The same booth, later. The technician is just finishing testing the
equipment as prescribed in the company manual.
SUPER TITLE: 1. TESTING—PRIMARY AND SECONDARY CIRCUIT

>   NARRATOR (VO)
>   Step one is testing. It involves running
>   all appropriate tests on the primary
>   and secondary circuits. With this
>   accomplished, we can move on to
>   DISSOLVE
INT BOOTH—DAY
Later still. The repair tech has his tools out and is soldering in a
new coil.
SUPER TITLE: 2. REPAIR—ON THE SPOT OR REFERRAL

> NARRATOR (VO)
>
> Step two is repair. This involves fixing
> the bad circuit. Repair can be done on
> the spot, if time permits, or by referral
> to a later date.

INT CUSTOMER'S OFFICE—DAY
She is on the phone with the tech getting word the Electrostat is
fixed. She is obviously pleased.
SUPER TITLE: 3. INFORM CUSTOMER—PHONE, MAIL, OR FAX

> NARRATOR (VO)
>
> Assuming we've done it on the spot,
> step three is next: informing the
> customer. This can be done by phone,
> mail, or fax.

INT BOOTH—DAY
ECU—Referral notice attached to an Electrostat.

> NARRATOR (VO)
>
> Obviously, step three is carried out
> later if repairs have been referred to a
> new date.

EXT HOSPITAL—DAY
At a company truck in the parking lot. The tech loads his tools
into a truck and drives off to another repair job.

> NARRATOR (VO)
>
> By following these three steps—testing,
> repairing, and informing the
> customer—you can be a successful
> Electrostat workstation repair
> technician.

Version one of this piece has, among other faults, major structural
problems. To begin with, the material has no introduction. We are thrust
into step one without reason or preparation.

The material is then presented to us out of order. In just one case,
step three follows step one. Before we have gotten oriented, we must
quickly reorient ourselves as to which step we are on and why we've
skipped to three instead of progressing logically to two. If this doesn't mix
things up enough, even *parts* of steps have been taken out of order and
mixed with other steps. In step three, there is a reference back to the
second part of step two, the referral process.

The first version also has no transitions whatsoever. We simply cut
from scene to scene leaving one idea and being hurried, it seems, straight
into the next.

On top of these structural problems, the general narration is worded
poorly, and the use of support titles is weak. The result of all this is a
jumble of ideas that would discourage anyone from even trying to pay
attention.

The second version has much more clarity. The content is first introduced with a brief overview. This sets us up for what's to come, allowing us to become mentally prepared to absorb this subject.

Although this version also makes better use of titles and is written more concisely, a good part of the reason for its clarity is simply the *arrangement* of the content (the test and repair steps) in the logical one, two, three order, as it would normally be carried out.

Finally, this version has a tidy little wrap-up, briefly summarizing what's been covered. It lets us know that what we've been discussing is now fully communicated and uses the moment to briefly reiterate the three steps.

Transitions are also provided in this second version. In this case they are dissolves and simple verbal cues. The dissolves tell us visually that time has passed between the shots that take place at the same location, and the verbal cues move us smoothly from idea to idea.

The introduction and step one, for instance, are bridged by "Let's start at the beginning." Steps one and two are bridged with "With this accomplished, we can move on. . . ." Between steps two and three, it's "Assuming we've done it on the spot. . . ." The result is a comfortable-to-follow, easily absorbed technical piece with a very high clarity factor and thus maximum instructional value.

Benefits Bookends

Another structure often used in corporate and sales programs is what I call the "benefits bookends" structure. It tends to be more effective with motivational/informational content than with heavily instructional or technical material.

The benefits of becoming involved in some company campaign are presented as an opening teaser, the front bookend. It might be something like this.

> FADE IN
>
> MONTAGE—A series of exciting SHOTS of people vacationing in areas offered in the company's "Sell-for-Sun" program: snorkeling, sunning on the beach, dinner on a beachside terrace, dancing by the surf at sunset.
>
>   HOST (VO)
>   Vacationing in paradise . . . Mexico . . .
>   Hawaii . . . the Caribbean. Getaways
>   only for the rich? Actually, they're
>   waiting in the sun for <u>you</u> . . . and in
>   the next few minutes, we'll show you
>   how to get there!

Following this, perhaps our program title is superimposed over an exotic shot. We might then make a transition with a video effect and music to an on-camera host who, with the help of titles and cutaway footage, would take us through the program body. The body would be made up of the steps to follow in order to win a cruise. Our host would cover them in much the same way we covered the Electrostat steps, in a logical, straightforward order with appropriate transitions.

A close for this program—the *back* bookend—might then be handled like this.

```
EXT LAGOON—DAY
The host has snorkeling gear on. He is standing under a palm tree
with a group of vacationers who are ready to go diving. He stops
his conversation with them just long enough to turn to the
camera and say . . .

                    HOST
          So, it's up to you. Spend your next
          vacation trimming the hedges, or sell
          for sun!

He turns and heads for the snorkeling trip. MUSIC UP.
                                        DISSOLVE
HIGH ANGLE—the lagoon from an overhanging cliff. The group,
including our host, wades in and begins snorkeling. SUPER
CLOSING CREDITS, MUSIC DOWN AND OUT, as we . . .
                                        FADE OUT
```

Contrasting Actions

Still another structure used in corporate programs is what I call the "contrasting actions" structure. This can be effective when communicating the right and wrong ways to carry out certain procedures.

First the audience is shown, for example, the wrong way to try and make a sale. A host or narrator makes note of the mistakes, then explains that sales can be closed more often by doing it her (the right) way. We then see the right way, and the correct steps are noted. The customer is left much happier, and the sale is closed.

Creative Use of Structure

Remember that the structure of your material and the transitions you use to bridge it are critical to the clarity factor and thus to the success of your script. Also remember, however, that the concept or story into which you place your structure need not be boring or "educational" sounding. I once used the "tell 'em" structure, for instance, as a content arrangement method for a humorous program on the proper steps for digging without causing back injuries.

The program concept centered around two grave robbers. One, Mr. Princeton, was the lazy and crafty brains behind the operation. The other, Barcaloo, was the huge, stupid assistant who just couldn't seem to get it right. As the two prepared to rob a grave and Barcaloo began to dig, Mr. Princeton provided the first part of the structure like this:

EXT GRAVEYARD—NIGHT
Barcaloo begins to dig. He hunches over, stretches out, and grabs
the shovel low on the shaft: all the wrong digging methods. Mr.
Princeton becomes infuriated.

                    MR. PRINCETON
          Barcaloo, you oversized tub of stupid
          cellulite! You're doing it all wrong!

                    BARCALOO
          Uh, . . . but, I. . . .

                    MR. PRINCETON
          But, shmut! Listen to me. If you are
          going to carry this casket back to the
          lab, and believe me you are going to
          carry the casket, your back must be
          undamaged! Do you hear?

                    BARCALOO
          Yeah, but. . . .

                    MR. PRINCETON
          Quiet! Just listen. There are three steps
          involved in digging correctly.
          Regardless of your microscopic mind, I
          believe they are easy enough for even
          you to follow. Now listen up.

At this point Mr. Princeton briefly ran through the three steps.

Following this, the body of the program covered the three steps
in complete detail, with demonstrations of each by Barcaloo and Mr.
Princeton. The wrap-up was provided by a constable who had been
eavesdropping on the entire affair and finally arrested the two. As he
dragged both grave robbers off to prison, he thanked Mr. Princeton for the
tips on digging—mentioning all three steps—and asked if they would also
work for breaking rocks with a sledgehammer!

As we've seen, then, provided your design calls for it and your
producer and client are amiable, you are free to be as creative as you like
within the parameters of whatever basic structure you choose. Your only
guideline should be assuring yourself that the information is communicated effectively.

■ ■ ■ ■ ■

You Script It 10

When you are working out program and segment concepts, it often helps to
consider visual analogies, which are elements or situations that offer a kind of
instructional or informational point of reference for the viewer.

One visual analogy used recently in a series of very effective substance abuse
commercials is the "this is your brain on drugs" idea. In it, as a spokesperson alludes
to a drug user's brain "on drugs," the camera tilts down to show two eggs frying in a

pan. The visual appearance of the eggs, the direct narration, and the greasy, crackling sound drive home the point in a visually graphic and very powerful way.

Another similar commercial uses a roller coaster ride as a visual analogy to symbolize the grip substance abuse can gain over drug users. In this commercial, the substance abuser, obviously frightened by a "bad trip," attempts to get off the roller coaster, but the safety bar is locked in place. Paralyzed with fear, the user-rider has no choice but to stay on for another "trip." Again, this commercial is visually tense, graphic, and hard-hitting.

In a corporate program designed to make employees aware that a change in mind-set had to occur in the organization very quickly, the analogy of a ticking clock was used as a segment concept. A close-up of a ticking clock was used to open and close the program on a dramatic note. For the close, the clock visual faded out and the sound of the ticking alone remained for several seconds over black.

What simple visual analogy might you use for a program discouraging dishonest behavior on the job?

PART III

The Scriptwriter's Work

In the following chapters you will find six examples that illustrate the skills we have explored throughout Part II. The first four examples, in Chapters 13 through 16, were parts of the same creative "package," which included a program needs analysis, a content outline, a treatment, and a script.

I chose these as examples for several reasons. First, they provide a good illustration of how all four steps work together in the development of a single program. In addition, the script itself illustrates the typical screenplay format. This project also involved developing a sales and motivational concept, which meant more emphasis could be placed on creativity than on content. Finally, the budget made possible developing a concept based on a fairly elaborate role-play situation that involved humor, characterization, motivation, and heavy use of dialogue. In other words, the project was fun.

Next is Chapter 17, which includes a complete script in a two-column format. It is a more informational and traditional type of corporate script with an on-camera host and brief vignettes illustrating the content points. It also includes documentary-type scriptwriting for a segment of testimonials.

The final example, in Chapter 18, is a short, two-column script developed entirely with stock footage. I've included this script because, in terms of production value and creative development, it is at the opposite end of the scale from the four-part role-play package. It was produced with virtually no money in very little time with almost no resources. It also offered very little creative opportunity but was nonetheless a worthwhile project.

With each example, I have provided personal commentary and insights into client, producer, and political factors, which had definite effects on the direction each project took.

A Program Needs Analysis

13

□ □ □ □ □

In earlier chapters we established that design research is one of the first and most important parts of the scriptwriting process. The following program needs analysis is one final form that type of research can take.

In the companies you write for, it may take a different form. It may be more or less extensive, getting into areas like in-depth cost analysis, visualization, or long- and short-term company goals. Then again, it may not even *exist* in the companies you write for. Your producer may ask you to submit a treatment and script and never even mention the subject of design research.

Whatever your situation, remember, regardless of the "deliverables" of your contract or the operating policies of the video department you are writing for, design research is an essential part of your work, and it should be carried out on every script you write.

As you read this program needs analysis, think or refer back to the design information covered in Chapter 4, such as objectives, audience analysis, and utilization. Ask yourself how these elements should affect the design of this program, and compare your ideas with what actually happened.

Win the Connection
A Program Needs Analysis

PROGRAM NEEDS ANALYSIS

PRODUCTION NUMBER	PP-8910
DATE	NOVEMBER 27, 1990
WORKING TITLE	WIN THE CONNECTION
SUBJECT	EMPLOYEE SALES REFERRAL PROGRAM
CLIENT	RITA WALLACE, SALES SUPPORT MGR.
	117 6TH, LOS ANGELES, CA 91301
	2908/500, 213/555-1456
OTHER APPROVALS	JILLY VERNA
	KLIEN & ASSOC., FOUR BOSTON FORUM,
	BOSTON, MASSACHUSETTS
	06904, 203/555-3872
	TIM ANDERS, PUB. AFFAIRS DIR.
	117 6TH, LOS ANGELES, CA 91301
	3110/500, 805/555-7150

CONTENT EXPERTS	JOHN JENNINGS, SALES SUPPORT MGR. 117 6TH, LOS ANGELES, CA 91301 2908B/500 555-6145
FYI COPIES	N/A
DEADLINE	February 24, 1991

COST/BENEFITS/IMPACT

The program outlined in this proposal will cost approximately $42,000 to produce. When completed it will be used as a motivational session starter for kickoff presentations of the Win the Connection Program in NuComm, Inc.

Win the Connection has the potential to produce millions of dollars in revenues for NuComm. Since the videotape program described in this proposal will give employees their very first exposure to Win the Connection, it will be a critical factor in determining how profitable the referral program may become.

page 2

PROBLEM
Employees in NuComm are not aware of the Win the Connection program or the exciting benefits it has to offer.

BACKGROUND
Win the Connection was originally implemented in NuComm as a previous referral program called Sell One Service. As most employees will remember, Sell One Service was canceled in California approximately 2 years ago. The concept was continued, however, in several other NuComm companies, and it is with the benefit of those experiences that Win the Connection is now being piloted in California.

RELATIONSHIP TO COMPANY GOALS
Since Win the Connection is a companywide program, providing the videotape to support it is in keeping with the company's goals. In addition, company goal #4 specifically calls for NuComm to "explore new methods of generating revenue," and Win the Connection certainly fits this definition.

AUDIENCE
Primary	all NuComm employees
Size	approx. 17,000

AUDIENCE DEMOGRAPHICS
Most audience members are experienced NuComm employees. All have high school educations and roughly 15% have college degrees. The male-female split is approximately 50–50 and the ethnic mix is typical for Southern California, mostly White, African-American, Hispanic, and Asian-Pacific employees.

AUDIENCE INTEREST/NEED
Employees interested in exciting prizes and luxury vacations will see an important need for the Win the Connection program. Other audience members will at least have an initial interest in Win the Connection.

Whether or not that interest turns into open support will probably depend on two factors: whether employees view the program as viable and whether they feel there is definite reward in it for them. This videotape program will attempt to capitalize on these two factors to excite and motivate employees.

AUDIENCE KNOWLEDGE/EXPERIENCE

Many audience members will have had brief contact with the Sell One Service program and thus will be familiar with its general workings. Details of Win the Connection will be very similar and will be provided by the live presenter. The videotape, however, may provide a basic overview.

AUDIENCE ATTITUDES

Audience attitude is a key factor in the success of Win the Connection. Sales referral programs similar to this have sometimes carried a negative image. This is primarily due to the fact that selling seems to be uncomfortable for many employees. In addition, the return for effort in some programs has been perceived as minimal.

For these reasons, the first impression employees get of Win the Connection should be positive, fun, and exciting. In addition, acknowledging the fact that Sell One Service was not successful but that we have learned from that experience should help establish credibility. The substantial prizes available through Win the Connection should also have a positive influence on audience attitudes, as should the simplicity and ease of the referral process.

OBJECTIVES:

Primary:

Having viewed the proposed program, audience members will:

1. be able to define Win the Connection as an employee sales referral program using an 800 telephone number system.
2. be aware that the potential for big winnings exists with Win the Connection
3. be motivated to hear more about the details of Win the Connection from the live presenter

NOTE: In addition to the stated objectives, top executive support for Win the Connection must also be displayed in this program.

DISTRIBUTION

The distribution format and number of copies will be discussed as production moves forward on this project.

UTILIZATION

All copies of this program will be utilized by local managers trained as live presenters.

The presenters will follow tape viewings with detailed handout information on the workings of Win the Connection and the paperwork and phone numbers needed to actually start making referrals.

EVALUATION

This program will be evaluated with the use of a brief questionnaire handed out by the live presenter.

Commentary

The initial information contained in this company's program-needs analysis format is a series of basic facts about the project, such as dates, project number, clients, and approvals. It is a part of the document, because the program needs analysis (PNA) is one of the handiest places to record this information for easy future reference.

The cost/benefits/impact is another important area taken into account on every video project produced in this company. The producers always require an initial budget figure and a projection of what that investment will "buy back" for the company.

In this case the budget figure was provided by the producer himself and the buy back is stated in very general terms that do not project a specific monetary return as a result of the tape. Instead, it mentions that the tape will be the first exposure employees will have to Win the Connection and therefore a critical factor in how much of those "millions of dollars in revenues" will result from their participation.

In some cases, a producer might require that a very specific cost/benefits/impact section be included. If so, an analysis must be done on how much it would cost employees to get the same information they will gain from the videotape in other ways, such as print material, live instructors, or audiotapes. A total cost for each other method is then compared against the cost of the videotape.

As an example, the management personnel running the Win the Connection program might have determined that another way to achieve the excitement they were after was with the use of a live performance of some sort. The cost of this performance would have included the cost of the performers' time and their meals, travel, hotel, and so on as they made their way around the company.

In addition to cost, other elements to be considered would have been time and effectiveness. With the videotape, for example, copies could be distributed to every work location, and, using locally drafted presenters, everyone could get the same message on the same day. In the case of traveling performances, the rollout or start-up of the project would require time (probably days or weeks) for the performers to visit each operation.

Whatever the math and other considerations, you can see that the reason for this part of the analysis is to be sure the cost of the videotape program is worth the investment and that there is not some cheaper way to get the same result.

The problem statement in this needs analysis is a single sentence stating the overall purpose of the program. As we've discussed, this problem-oriented way of stating the purpose establishes the need for the videotape to provide a solution to the problem.

An interesting political factor emerged in developing the background section of this needs analysis. The program had been previously tried in the company 2 years before under another name, Sell One Service. This first attempt had failed and left a bitter taste in employees' mouths.

This type of information would be important in the development of any program on a subject like this because it could have an effect on audience perceptions. As it turned out, the client wanted it included in the needs analysis document, but she made it clear she *didn't* want it as a point of focus. Instead, she wanted to acknowledge the old program briefly and then get to the positive business of kicking off the new program. Acknowledge the negative but *don't* dwell on it is a common way many corporate programs deal with politically uncomfortable or controversial issues.

As it turns out, although it was important information, this project (luckily) took a direction that didn't even need to acknowledge Sell One Service.

As you can see, the audience size for this program was large. This is often another factor in determining the amount of money that should be invested in a project. Had the total audience size been 100 instead of 17,000, the potential revenues generated by Win the Connection would have been well below the "millions" mark. In that case, I doubt we would have gotten the budget to do the show as it was eventually written.

Audience demographics, interest, and attitudes explore some of the critical human issues that eventually helped dictate the direction this program took. In working through this information and the objectives section, it became clear even as we began the project that simply making the new Win the Connection appear to be fun, exciting, and profitable could get it off to a successful start.

The objectives in this needs analysis were each developed for important reasons. First, the client felt that we didn't have to impart a great deal of information (that would be done by a presenter and written material), but that the viewers should at least know what type of sales program we were presenting, generally how it worked, and the basis for calling in referrals.

The second objective focuses on the "what's-in-it-for-me" aspect of the program. This was another important element because in this case the prizes turned out to be pretty impressive.

Finally came what developed into the key objective. We all felt that if the tape accomplished nothing except genuine employee interest and excitement in the Win the Connection program, it would be successful.

As you can see, the first of these objectives is instructional. It is stated in terms of a specific viewer behavior: being able to define the Win the Connection program. The second and third objectives are motivational. Although the expected viewer behavior is stated in a less specific way — "be aware" and "be motivated" — the motivation element was actually the key to the success of the project.

Of the final sections in this analysis, the utilization is most important. The fact that a live presenter was to accompany each showing and provide detailed written information coincided perfectly with the objectives and audience analysis. It meant we could keep the program short and write the script primarily as a motivational piece that included almost no "hard" content. Our project could be totally exciting, and we could leave the boring stuff for the presenter and the written handouts.

14

□ □ □ □ □

A Content Outline

In our chapter on content research, we discussed the idea that writers generate different kinds of content outlines based on their own writing styles and the requirements of each project. As you may recall, we concluded that the requirement to develop a formal outline was more likely on technical projects requiring in-depth research. This, we said, was because technical scripts also tended to require more organization and structure—the very things a content outline provides. There are exceptions, however, to this unspoken rule of thumb, and this turned out to be one such case.

The following content outline was requested by the client on the Win the Connection project. In most cases, such a project would not require a formal outline; even in its early stages it was emerging as more of a motivational program than a content-based project. With development moving in this direction, an informal outline, briefly highlighting the content and setting up a good solid structure, would probably have been sufficient. In this case, however, the client wanted to see a content outline as part of her development package for several valid reasons.

First, she really didn't know at this point (nor did I and the producer) exactly what direction the program would take.

Second, this was a very important project for this particular client. She had been given the job of reinspiring a sales program that had previously failed. She was determined to make it work, and she wanted every base covered as thoroughly as possible as she implemented her plans—including being sure the content of the videotape was accurate and well organized.

Finally, the client's personality required detail and orderliness. She probably would have wanted a content outline written whether the project was critical or not and regardless of the type of show it was to become, just because she was that type of manager.

As content outlines go, this one is not extensively detailed. At places, for instance, it refers to other documents such as company bulletins for additional details. What it does contain, however, was more than enough for this particular project. You will no doubt experience some assignments in which more detail will be required and some requiring less. Like all other aspects of this book and the writing field in general, what is effective in a specific situation is usually more important than a rigid format or requirement.

As you read over this content outline, try to recall our discussions on content research and be most cognizant of the way the information is structured. Also consider which of the facts were considered unimportant enough to have been left to other documents and which were considered necessary for inclusion here.

WIN THE CONNECTION
A Videotape Program Content Outline

I. INTRODUCTION

Win the Connection is an exciting employee sales referral program now being piloted in NuComm, Inc. The program includes an impressive list of valuable prizes, including jewelry, home electronics, tools, furniture, clothing, luxury vacations, and five 10-minute shopping sprees in a San Francisco warehouse packed with nearly $9 million in prizes!

II. HOW IT WORKS

''Sales by employee referral'' is the key phrase in describing Win the Connection.

Employees of NuComm, Inc., have an immense network of friends, neighbors, and on-the-job contacts. Many times, though an employee may not realize it, one or more of these contacts has a communication problem that might be solved by a NuComm product or service.

Employees involved in the Win the Connection program help solve these problems and win exciting prizes at the same time. This is accomplished by mentioning the products and services NuComm offers to the many members of an employee's network of acquaintances.

If the acquaintance would like to hear more about the product or service, the employee makes a simple referral to the Win the Connection headquarters by means of a toll-free 800 telephone number.

When a Win the Connection referral results in a sale, the employee who made the referral is awarded points. Total points acquired then give the employee a number of possible ways to win valuable prizes.

The Win the Connection program will be conducted in ongoing 6-month phases. Phase one begins on March 1, 1991 and ends on September 1, 1991.

III. WAYS TO WIN

There are three ways an employee can win standard prizes at Win the Connection: first three to 25,000 points, monthly drawings, and total acquired points.

1. First Three to 25,000 points
The first three employees to reach a 25,000-point total each month will be automatic two-way winners. Each of the three employees will:
A. win any 25,000-point prize shown in the Win the Connection catalog
B. be included in ongoing monthly ''First to 25,000'' drawings. Drawings will begin when this milestone is reached by the first 10 employees

2. Monthly Drawings
There will be monthly drawings on an ongoing basis throughout the Win the Connection program. All employees with any sales during that month will be included in these drawings. Point totals have no bearing on the monthly drawings.

3. Total Points Acquired
Employees will be eligible to receive prizes of their choice based on total points acquired. Prize point values have been calculated based on list prices and are shown in the Win the Connection catalog. Prize value

categories are based on an ascending 1000-point-per-segment structure. There are 100 segments in all, making the top prizes worth 100,000 points.

IV. WHO IS ELIGIBLE

All employees of NuComm, Inc., may participate in the Win the Connection program.

V. POINT VALUES

An earned point value system has been developed for Win the Connection. The earned point value means the amount of points acquired by the employee for the sale of a particular NuComm product or service.

page 3

Earned point values have been based on two factors.

A. Total retail value of the product sold
This includes products that are direct sale items and produce no monthly revenues.

B. Retail value plus monthly 6-months revenue
This includes both products and services that have an initial retail value and earn a monthly revenue in addition.

Note: Exact point values for each NuComm product are shown in the Win the Connection catalog and Division/Staff Bulletin #S-348.

VI. PRIZES/AWARDS

All prizes available in the Win the Connection program are shown in the Win the Connection Catalog and Division/Staff Bulletin #S-348.

The general prize categories are

> home electronics
> tools
> furniture
> family fashions
> jewelry
> hardware

Additional prize categories include luxury vacations to

- the Caribbean
- the Orient
- Alaska
- the Mediterranean

The grand prizes are
five 10-minute warehouse run-throughs.

Note: A warehouse run-through is a 10-minute free run through the Klien & Associates awards warehouse in San Francisco, California. The five grand prize winners will receive a 2 x 5 x 3-foot loading cart and an assistant to help load items. Details are contained in Division/Staff Bulletin #S-348.

Employees will receive company-paid transportation to and from the warehouse and all expenses for themselves and their spouses.

VII. SUMMARY

The Win the Connection employee sales referral program is an exciting
and profitable way for employees to help NuComm, Inc., retain its
dominance in the telecommunications marketplace.

Employees need only mention NuComm products and services to friends,
acquaintances, and customers. A simple 800-number referral process then
follows. Resulting sales have the potential to earn employees prizes such
as home furnishings, entire wardrobes, or exotic vacations. Grand prize
winners will be treated to an all-expenses-paid 10-minute shopping spree
in the Klien & Associates warehouse in San Francisco, California.

Commentary

The basic structure chosen for this content outline was a motivational
version of the benefits bookends. It begins with an introduction that is
actually a "hook" overview of the Win the Connection program. This
establishes the subject in the reader's mind and provides motivation to
learn more.

Following the introduction comes the body of the outline, five
sections centered around a series of key content points. Finally, the piece
is wrapped up with a summary that makes a last pitch for the benefits of
becoming involved in the Win the Connection program.

I decided that the body of the outline should present the factual
aspects of Win the Connection in an order based on what key audience
questions or reservations might be about the program: How Win the
Connection worked, ways to win, who could be involved, how the point
system was set up, and, the old standard — "what's in it for me" — the
prizes. These key audience questions were acquired through audience
and client interviews during the content research phase of the project. I
was comfortable with this type of presentation because I felt that if the
program did become an informational one, the type and order of the
information could then be lifted almost straight out of the content outline for inclusion in the treatment and script.

As it turns out, the possibility of an informational program, was
wasted anxiety on my part. The outline was quickly approved by the
producer and client, and I then moved ahead into what became a very
*non*informational treatment.

15

□ □ □ □ □

A Program Treatment

As you might expect, conceptualizing and visualizing this program was a critical phase in its development. As the producer, the client, and I met over time to discuss the needs analysis and content outline, the same key words kept coming up: "fun and exciting."

Because a presenter would be there with ample written material and a live verbal presentation would follow every showing of the tape, the amount of content we had to include was minimal. Also, because Sell One Service had given the audience something of a sour taste, we felt we needed either to honestly address it as a failure or to do something that would take the audience's attention completely away from it.

As for the requirement of displaying top executive support, we were all pretty well agreed that the company president would do a typical talking head open or close to the program. From his office, he would briefly state his personal support for Win the Connection.

In the final content meeting before I left to write the treatment, we had reached a consensus: if we could leave the audience laughing and feeling excited about Win the Connection, the videotape would be a complete success. In addition, two other criteria had been set: inclusion of the executive shot just mentioned and a series of shots of an actual run through the warehouse, which had been sent to us from Klien & Associates. This run-through footage was so fast paced and exciting that we all felt a portion of it should somehow be included in our program.

With these ideas uppermost in my mind, as well as one other key element that we will discuss a little later, I went to work. During the next week, I developed the treatment that follows.

---

page 1

WIN THE CONNECTION
A Program Treatment

As our program opens, we are amazed to see Allan Casey, the NuComm vice president known to us all, being hauled into a seedy downtown police department. As Casey is led through a cluttered office booking area in handcuffs, he looks tired and dejected. We see, from a distance, that he appears to be pleading for his release.

From an adjacent office, Police Sergeant Davis, a grumpy, hardened, 30-year veteran of the force, and the arresting officer, Detective Parks, a chubby man, short on brains, look on.

Davis asks Parks for the scoop on this arrest. Parks replies that Casey was picked up driving through town with five TVs, three stereos, and assorted other equipment in the back of his pickup truck. Robbery is suspected, of course, and Parks says Casey's alibi was quite bizarre.

As Davis chuckles at the absurdity of what he now hears, Parks goes on to tell him that Casey claims he is a vice president for a large communications company, NuComm, Inc. In addition, he swears he won all the equipment found in his truck through a company sales referral program called Win the Connection.

"Ha! Win the Connection. Sure thing," Davis says. "What does this guy think, we just fell off the turnip truck?"

Parks replies, "I know it sounds crazy, Sarge, but the guy is sticking to his story. He makes this Win the Connection sound like the greatest thing since electronic cell doors."

Although Davis is still skeptical, he asks to hear more. Parks now gets out his small, black notebook and begins to leaf through it, relaying what he has been told by Casey.

---

page 2

Win the Connection, he says, is an employee sales referral program with an 800 call-in number and an incredible prize package. Basically, employees talk with friends, neighbors, or people they meet on the job about services and equipment the company offers: custom calling features, intercoms, telephones, dialers, and so on. If one of an employee's contacts decides he or she wants more information, he or she can arrange to get it through the employee. He or she simply takes the appropriate information down on a company-provided form and calls it in to the convenient 800 number. When the order is actually sold, the employee gets points based on a sliding scale value system: the more valuable the sale, the more points acquired. Later, the employee can redeem the points for prizes displayed in a handsome catalog collection of items, everything from jewelry to small sailboats.

"Sailboats!" Davis exclaims.

"Yeah," Parks continues, "and that's not the best of it. They got luxury vacations, and the five top point earners, like that Casey guy, if he's on the level, that is, get what they call the 'run through the warehouse.' "

Davis chuckles and shakes his head in further disbelief, but he has Parks continue. Parks tells him the run through the warehouse is the grand prize. The five top-earning employees are given a 10-minute shopping spree in a warehouse full of incredible things like TVs, stereos, furniture, CD players, and much more. They're given a huge storage tub to roll around and a cheerleader to follow and even a guy to help load whatever the employee wants!

Davis then makes the assumption that it's this run through the warehouse that this Casey character claims to have used to fill up his pickup truck with the TVs and stereos.

Parks confirms this. He goes on to say, "And here's the kicker, Sarge. He's claiming his alibi can be backed up by his boss, the president of NuComm, John Crain. He also says the lady who actually runs the Win the Connection program, Rita Wallace, will back him up, too!"

When Davis hears this, his face lights up. He's sure he now has Casey cornered in the lie that will expose his entire bizarre scam. "Well, then," he says, "we'll just take him up on that. We'll call the president of NuComm right now and just see what he has to say!" With that, he dials the phone and calls John Crain's office.

---

page 3

As it turns out, when the call comes in, John Crain and Rita Wallace are seated in Crain's office discussing the recent results of the Win the

Connection program. Crain's secretary passes the call on to him, and Sergeant Davis briefly explains the situation along with the alibi given by Casey. He does not, however, mention Casey's name yet.

When Mr. Crain hears this story, he realizes it's probably a legitimate call, but we can see that a light bulb has clicked on. He suddenly gets a fiendish idea. "Ah, and just what is the, ah, name of this person claiming to be a NuComm vice president, Sergeant?" he asks.

"Casey," Davis replies, "Allan Casey. Does it ring a bell?"

Crain looks at Wallace, covers the receiver, and says, "You'll have to excuse me, Rita, but I've been waiting to do this for about 30 years now." He then replies to the sergeant, "Sorry, Sergeant, but actually the name doesn't ring a bell."

The sergeant chuckles and nods, thinking he has been right all along. Allan Casey is indeed a crook.

As our program nears its close, Casey is seen under the lights in an interrogation room with the sergeant. "I swear!" he is pleading. "It's true! I've worked there for 30 years! I can't understand this! Did you tell him my name? There must be a mistake!"

The following message then appears over this shot of Mr. Casey pleading.

> John Casey was sentenced to 20 years of hard labor, loading prizes for NuComm employees at the Win the Connection warehouse.

As this picture fades away, we close the program with a series of exciting shots from videotapes of the first warehouse run-throughs. These were taped on September 16 by Klien & Associates. Master duplication copies are being sent to NuComm.

Commentary

The additional key element I alluded to earlier was the reputation and personality of NuComm vice president Allan Casey. Although he was the top executive under the president, his personality was hardly a stuffy or conservative one. On the contrary, he was known as a likable prankster, an excellent impromptu speaker, and often the executive comedian. He had also earned great respect as a manager. On top of all this, he had come up through the ranks and was widely known, even by frontline craft employees. These qualities eventually made him the perfect choice to star in the Win the Connection videotape.

Before arriving at that idea, however, I spent a good deal of time brainstorming concepts for this program. I remember sitting at a local fast-food restaurant going through coffee, Cokes, and french fries and writing things like

Content	Concept	CDP
Lots of great prizes	• Host in warehouse surrounded by prizes	Too exp — warehouse in Frisco. Fake? Set?
	• Fast montage	Yes
	• Run through warehouse footage	Yes
Anyone can win big	• Regular Joe gets entered in drawing	Could wrk Fake Caribb?

	and ends up in Caribbean under a palm tree	Back drop? Comedy?
Executive support	• Exec in office Brief support statement	Okay — Who? Crain? Casey? Open w/hook Do execs as *close*
How it works	• Employee goes through process; vignette	Okay, may not be exciting
Prize values	• Titles over pictures of prizes	Okay
Old version failed	• Host — straight to cam, short.	Yes, direct No pulled punches Then get off it

While I was finding ideas that I felt would generally work, they all seemed too standard. I was also having trouble finding a way to work in the run-through-the-warehouse stock footage. In general, the sense of fun and excitement we were after just didn't seem to be present in any of the concepts.

I kept at it, however, for 2 days. While I was considering who should do the executive talking head, the idea of using Casey as on-camera talent occurred to me. He was funny, everyone knew him, and using him might just be the way to incorporate that executive support into the program itself, rather than adding the usual talking head appendage.

Despite all these positive aspects, at first I dispelled the notion. I was well aware that using executives as actors usually meant poor performances and thus disastrous programs. There was also the problem of getting his time and even his agreement to do it.

The more I brainstormed, however, the more the idea kept occurring to me. If I could make his involvement minimal, I reasoned, he might agree to do it. If any executive would agree, in fact, Casey was the one. With very few lines, he wouldn't have to act much, and his time away from the office could be minimal, especially if it were carefully scheduled.

After a good deal of wrestling with this and other possibilities, the concept of Casey being thrown in jail after being caught with a load of run-through-the-warehouse prizes emerged. This seemed funny enough in itself, but the kicker was the idea of his boss, President Crain, finally being able to get even with him for all the pranks he had pulled over the years. Audience members, all of whom knew of Casey's reputation and his relationship with Crain, would identify with it instantly. The two police characters also seemed like the perfect duo to get across the basic content of Win the Connection in humorous, naturally motivated scenes. On top of this, the overall idea seemed funny enough to overshadow the prior Sell One Service failure without even having to mention it.

I felt it had everything. Not only was it funny and exciting, it was also unique, and it showed executive support simply by their involvement in the production of the program.

The only real drawbacks seemed to be executive agreement and acting ability, but, again, I reasoned that if their parts were kept minimal it could be pulled off.

Because the idea was fairly radical, however, rather than write a complete treatment, I first worked out the details, briefly listed the story points on a sheet of paper, and called the producer. I took him through the story concept and voiced my feelings on why I felt it was the way to go. After hearing me out, he said he loved it and gave me the go-ahead. I went to work at once, carefully visualizing each scene, and had the treatment done in a few days.

When it was complete, I presented it to the producer. He read it and requested a few minor changes. With the revised version submitted the following day, the producer called a client meeting, and we all sat down in a conference room.

The producer presented the treatment to the group very skillfully. He first made it clear that the idea was not what he usually produced, but he was sure it was the right way to do this particular show. He went on to say that he would explain why he felt so strongly about it after presenting the story. He then handed out copies of the treatment and read it out loud. He followed this by presenting everyone with a copy of the needs analysis they had previously approved. Using it as a reference, he reminded the clients of the purpose for the program and the objectives we had developed: to generate fun and excitement, to include minimal content, to display executive support, to motivate employees to become involved, and, as we had all agreed, to leave them laughing.

After his presentation, everyone was convinced we had hit the nail on the head. Full approval was given, and I was on to the script.

A Screenplay Format Script

The script for Win the Connection came very easily. I wrote it over a period of 4 days. By the time I sat down to write it, I had been over most of the visualization in my head until it was crystal clear.

I could "see" very vividly the interior of the police station, the two officers, and the company president sitting in his office saying he'd never heard of Allan Casey. The other details were equally as clear and thus easy to convey.

As you review the script, consider all aspects of the screenplay format we've discussed in previous chapters. Look at the way the script is laid out on the page and the amount of scene description and the technical terms that have been included (and excluded). Is the script written in master scenes or using individual camera directions? Was the choice effective? Ask yourself if there is believable motivation for every word spoken and each action described. Does the pacing seem comfortable? Are the characters credible and humorous?

Above all, simply ask yourself if the script *works*. Given the development scenario you've been taken through, do you feel it's the right way to have produced the program? Do you feel it would accomplish the objectives outlined in the needs analysis? Maybe most important of all, if *you* had been the writer, how would you have handled this project?

---

page 1

                WIN THE CONNECTION
            A Screenplay Format Program Script

                                              FADE IN

INT POLICE STATION—DAY
WIDE SHOT—TO ESTABLISH A typical, seedy, downtown police
station booking area: lots of clutter, dirty coffee cups, messy
desks, and groups of overworked, gun-toting detectives busy at
work.

An office, separated from this area by a glass partition, can be
seen at SCREEN RIGHT. CAMERA PUSHES IN to this area.

                                              DISSOLVE

INT SERGEANT'S OFFICE—DAY
A chubby POLICE DETECTIVE named PARKS is just coming into
his sergeant's office to present the facts on a recent bust. The
SERGEANT, a squat, harried man named DAVIS, is seated behind
the desk, tie loose, busily doing paperwork. When he sees the

detective enter, he stops and looks up. Parks hangs up his jacket and pulls a black notebook out of his shirt pocket.

NOTE: During the following dialogue, SEVERAL POLICE ASSISTANTS can be seen in the BG through the glass partition. They are carrying in large boxes: TVs, stereos, tools, etc.

> SERGEANT
> Okay, so what's the scoop on this guy you just picked up?

> DETECTIVE
> Pretty bizarre. Guy says he's from the phone company.

> SERGEANT
> The what?

> DETECTIVE
> Phone company. You know, NuComm.

> SERGEANT
> Yeah, well, he didn't look like no phone guy to me when they took him through a few minutes ago.

page 2

> DETECTIVE
> Me neither, Sarge. Looks a little too, ah, greasy to be a phone guy. But that's what he says.

The sergeant chuckles. He gets up, moves to a nearby shelf, and pours a cup of coffee.

> SERGEANT
> And pray tell what exactly is a phone company individual doing driving around town with a truckload of TVs and stereos?

> DETECTIVE
> That's what I asked 'im.

> SERGEANT
> And?

> DETECTIVE
> Win the Connection. . . .

> SERGEANT
> (confused, flustered)
> Win the what? You dummy, win what connection? I said, what's a phone guy doing with a truckload of TVs and stereos?

                    DETECTIVE
          No, no, that's what he told <u>me</u>, Sarge!
          It's some company program called Win
          the Connection.

The sergeant moves back to his seat.

                    SERGEANT
               (again, a chuckle)
          Yeah, sure thing. Win the connection
          with your fence, and if you get nabbed,
          tell the cops you're from the phone
          company.

                    DETECTIVE
          He says it's on the level. Says their
          people can win lots of prizes or
          something.

                    SERGEANT
          Truckloads of TVs and stereos? C'mon,
          Parks, when did you fall off the turnip
          truck?

                    DETECTIVE
          Sounds bizarre, I know, but at least
          <u>part</u> of his story checks out.

                    SERGEANT
          What do you mean?

                    DETECTIVE
          Well, we made some calls. Seems the
          same kind o' thing's been going on in
          some other states, and those people've
          all checked out legit. NuComm folks
          who hit it big on this same kind of
          referral program thing.

This throws the sergeant a bit of a curve. He pauses, thinks, puts
his coffee cup down, then leans forward.

                    SERGEANT
          Exactly what is this Win the
          Connection again?

The detective now refers point by point, as if reading a script,
from his notebook.

                    DETECTIVE
          Let's see here. It's an employee referral
          program. Employees "win the
          connection" with friends and
          neighbors, that kind of thing. They tell

people about products and services and
make a call to a special 800 number.
Then the salespeople call back. If they
actually make the sale, the guy who
first made the referral gets points.

                    SERGEANT
And?

                    DETECTIVE
And if he gets enough, I guess he gets a
10-minute free run through this huge
warehouse where they got all this stuff.

                    SERGEANT
And he gets anything he wants?!

                    DETECTIVE
Anything! Stereos, lawn mowers, CD
players, TVs, you name it.

                    SERGEANT
And what if he doesn't, ah, ''hit the big
time''?

                    DETECTIVE
Well, I guess there's other stuff too.
Watches, tools, gold jewelry, all kinds
o' goodies.

The sergeant is now realizing this all makes a certain amount of
sense, but he's still not sure how much.

                    SERGEANT
                    (pondering)
Hmmm. Sounds legit, I guess. But, you
know, I just get a funny feeling about
this guy, I don't know. It's just
something about the way he came
strollin' in here with all those TVs and
cameras and remote controllers and
stuff stickin' out of his pockets.

                    DETECTIVE
I felt the same way, Sarge. Gut hunch
or somethin'.

                    SERGEANT
Can we check him out?

                    DETECTIVE
Well, he gave me this number. Says it's
the president's office.

> SERGEANT
> The <u>company president?</u>

A light bulb is now clicking on for the sergeant.

> DETECTIVE
> Right. He says all I have to do is call
> this guy, or some woman . . .
> (checks the notepad)
> named . . . Rita Wallace. I guess she's
> running this Win the Connection. He
> says either one'll verify his alibi.

The sergeant chuckles and snatches the detective's notepad. He then sits down and reaches for the phone. He's got it all figured out.

> SERGEANT
> Well, what do you say we just take 'im
> up on his little game. We'll just give
> this president a call! Here, where's that
> phone number?

Parks points out the number and sits down.

> DETECTIVE
> Second page, Sarge. Right there.

> SERGEANT
> (dialing)
> The president. Ha! Fat chance!

INT JOHN CRAIN'S OFFICE—DAY
JOHN CRAIN and RITA WALLACE are seated at a small
conference table discussing Win the Connection. Suddenly, Crain's
SECRETARY'S VOICE comes over the SPEAKER PHONE.

> SECRETARY
> Mr. Crain?

> CRAIN
> Yes, Anne.

> SECRETARY
> I have the police on line 2. They say
> they need to speak with you right
> away.

Crain looks at Wallace with concern.

> CRAIN
> Excuse me, Rita. This sounds
> important.
> (to secretary)
> Put them through, Anne.

NOTE: We now INTERCUT between the sergeant in his office and Crain and Wallace, as appropriate.

                    CRAIN
Hello.

                    SERGEANT
Ah, hello. Ah, Mr. John Crain?

                    CRAIN
Speaking.

                    SERGEANT
Ah, yes, well, ah, Mr. Crain, this is
Sergeant Davis down at the PD. We
have a, ah, suspect here, who claims to
be—get this now—an employee of
NuComm.

                    CRAIN
I see. So how can I help you, Sergeant?

                    SERGEANT
Well, ya see, sir, we caught this guy
driving down the street with a
truckload o' televisions, stereos, tools,
stuff like that.

                    CRAIN
I see.

                    SERGEANT
Right. And you see he claims he won
all this stuff in some program called,
ah, Win the Connection or something
like that. And
(chuckle)
. . . are you ready for this one? He says
all we have to do to verify his alibi is
give you a call, the president!

---

page 7

Crain now realizes there's probably been a mix-up. He smiles at
Wallace. She too smiles.

                    CRAIN
Well, Sergeant, I happen to have Ms.
Rita Wallace, the person in charge of
Win the Connection, sitting in my office
right now.

                    SERGEANT
So you do actually have a Win the
Connection program?

                    WALLACE
Yes, Sergeant, we certainly do.

                    SERGEANT
      Well, then, how about this character
      here? We're not quite sure about him,
      either. I mean, calling the president for
      an alibi?

                    CRAIN
      Tell me, Sergeant, what's this person's
      name?

The sergeant refers to his notebook.

                    SERGEANT
      Let's see here. Says here his name is,
      ah, Casey, Allan Casey.

Crain and Wallace look at each other with surprise at first. Then
Crain has an idea. He covers up the speaker and WHISPERS to
Wallace.

                    CRAIN
      Rita, you'll have to forgive me, but I've
      been waiting to do this for about 15
      years.

He then uncovers the speaker and talks to the sergeant.

                    CRAIN (CONT)
      Ah, Casey. Gee, I'm sorry, Sergeant,
      but the name doesn't ring a bell.

We hold on Crain as he and Wallace chuckle quietly, then . . .
                                        DISSOLVE

INT POLICE INTERROGATION ROOM—DAY   *page 8*
ALLAN CASEY is seated across from a POLICE INTERROGATOR.
The spotlight is on. Casey is sweating and pleading. The
interrogator isn't buying his story.

                    CASEY
      . . . But it's the truth! I swear! I've
      worked for NuComm for 34 years! I'm
      a vice president!

                    INTERROGATOR
      Right. Sure thing. And they've got this
      warehouse full of TVs and stuff for you
      people, huh?

                    CASEY
      Yes! They do! And when people win
      they get to . . .

As Casey now continues, VOICE SLIPPING UNDER, we . . .
                                        DISSOLVE

STOCK
Selected footage of the warehouse run-through.

> CASEY (VO)
> . . . walk through first, and then they
> get this big, ah, dipsy-dumpster, sort
> of, and they get to run up and down
> the aisle for. . . .

As Casey's VOICE CROSS-FADES TO MUSIC, the following TITLE IS
SUPERED:

John Casey was sentenced to 40 years of warehouse labor. He
now loads prizes and awards for employees cashing in on Win the
Connection.

> SUPER CLOSING LOGO. MUSIC STING AND OUT
>
> FADE OUT

Commentary

As you may have noticed, the only real change that came about between
the treatment and this script was the decision to have Casey revealed
near the end of the program instead of up front. We decided this for two
reasons. The first was Casey's time. To have him involved in a fairly
complicated scene such as the program's opening might make for several
hours of laborious shooting. These were hours the client and producer
felt he could better spend on his own matters.

Second, it seemed even better to reveal Casey as a surprise during
the later part of the show. This offered the viewer two surprises instead of
one. The first was the revelation that Casey himself was the person
behind bars. The second was Crain's decision to say he'd never heard of
Casey. We were convinced both would produce plenty of laughs.

Having read through this script once, you might reread it and think
or refer back to our discussion on dialogue, structure, and transitions. Are
those elements used as we had discussed? If so, did their use make for
what you consider an effective script? If the answer is no, ask yourself
why and what you might have done differently. Remember also that
there are good and bad concepts, but no concept is perfect. Maybe with
some thought you can come up with an even more humorous or exciting
approach. If so, write it as an exercise.

A Two-Column Script

"The Satisfaction Stealers" is a traditional instructional and motivational customer-handling script. Its primary purpose was to *teach* four basic skills involved in dealing with irate customers to bank tellers in a large company. The script was also designed to *motivate* these employees to use the skills in their day-to-day customer interactions.

I chose to include "The Satisfaction Stealers" as one of our examples for a number of reasons. First, it is a traditional example of the standard two-column format used extensively in corporate television today. Second, it incorporates a variety of segment concepts all interwoven into a single communication package. Those elements are an on-camera host, role-play vignettes with support titles, and employee testimonials.

A summarized audience analysis and the objectives established for "The Satisfaction Stealers" were as follows.

Audience
Craft-level customer service associates (tellers) in a large company—approximately 2700.

Demographics
Audience members are approximately 80% female. They are primarily high school educated, but roughly 10% have college degrees. Most are in the middle-income bracket.

Interest
Audience members should find interesting and unique the concept that irate customers "steal" their job satisfaction. They should also be interested in any techniques that can help them achieve greater job satisfaction.

Knowledge
Audience seniority levels average approximately 7 years. Most viewers know their jobs well and have used some form of customer-handling skills in the past. In many cases those skills have been handy hints picked up on the job or passed on by supervisors.

Attitudes
Attitudes should generally be positive, with some basic animosity toward hard-to-satisfy customers. Audience members should also be positive and attentive about learning new skills, and they appear to feel positive about the company in general.

Objectives

Having viewed the proposed program, audience members will

1. be able to state four primary customer-handling skills
   a. Let the customer vent frustrations.
   b. Always show sincere concern.
   c. Focus on the facts, not the anger.
   d. Always do your best to satisfy the customer's need.
2. be able to define the relationship between job satisfaction and customer handling: satisfying customers results in improved job satisfaction for individual employees
3. be motivated to try to satisfy irate customers in an attempt to gain more personal job satisfaction

As you read over the script, ask yourself if these objectives were met. Also, compare this format with the screenplay format, and consider the pros and cons of both from your own perspective.

---

page 1

THE SATISFACTION STEALERS
A Two-Column Program Script

FADE IN INT HOUSE—DAY AL MASON, a middle-aged man, is looking over his bank statement at the kitchen table. He's obviously upset at what he sees.	
	HOST (VO) Al Mason. His friends would call him just a regular guy. Right now, he'd probably call him<u>self</u> a victim of "the system," with a screwed-up bank statement.
With a determined look on his face, Mason folds up his paper-work and exits the SHOT.	If you happen to be the CSA he's about to visit, though,
DISSOLVE EXT HOUSE—DAY Mason's car backs out of the driveway and heads down the street.	. . . people like Al have a different name.
DISSOLVE EXT BANK—DAY ESTABLISHING SHOT as Mason pulls up in the parking lot.	They're called satisfaction stealers.
(MUSIC UP - LIGHT WITH A DRAMATIC EDGE)	

SUPER TITLE:	
SATISFACTION STEALERS: How to Beat Them at Their Own Game! LOSE TITLE—DISSOLVE TO:	
INT BANK—DAY A typical branch office. Customers and employees come and go. At the counter is a short line of customers. The HOST steps into the FRAME. She addresses CAMERA.	(MUSIC UNDER)

	HOST Just what are "satisfaction stealers?" And why would you want to beat them at their own game? Think about your job dealing with customers.
Gestures to customers behind her.	Do you get satisfaction from it? Do you go home at night feeling positive about what you've accomplished that day? If so, a lot of that satisfaction has come from dealing successfully with people because that's a major part of your work. On the other hand, if you're going home <u>not</u> feeling very positive, there's a good chance the reason is people like Al, people who, because of their anger or stubbornness, have gotten you upset at them and often at yourself. And, who knows, maybe even in trouble with your boss! Obviously, then, it stands to reason the less heartburn you let them cause you, the more job satisfaction you go home with and the happier you are at work.
CAMERA PANS during the following narration, leaving the host and REVEALING Al Mason approaching the front of the CSA line in the BG	
	HOST (VO) Now your next question might be, "Just how do I go about beating satisfaction stealers at their own game?" To answer that question, let's rejoin Al Mason, and watch an experienced CSA in

Melissa Wells, Second Street branch.

---

page 4

ANOTHER ANGLE—CLOSER
WELLS is just finishing with one customer, and Mason is next.

> WELLS
> Thank you, Mr. Gonzales. See you next week.

Gonzales nods, smiles, and leaves. Mason steps up.

> WELLS
> Hi, Mr. Mason. What can I do for you today?

Mason now starts off with a long, angry spiel while spreading out his statement on the counter.

> MASON
> What can you do for me? I'll tell you what. You can take that stupid computer of yours and deep-six it, either that or fire the person running it! Ever since you people installed it, I haven't gotten a correct statement!

---

page 5

As Mason's spiel continues, we see that Wells looks genuinely concerned, but she's also staying cool, calm, and collected.

> (DIALOGUE UNDER AND OUT)
> HOST (VO)
> Because she's experienced at handling people, Melissa knows she should first let an angry customer vent his or her frustrations, without interrupting. Why? Because at this point the customer wants nothing more than to simply get the problem off his chest. And interrupting that process is almost sure to bring on an even worse blowup. Once Al has gotten his frustrations out completely, Melissa knows exactly what to do as step two. Empathize and show a sincere interest in trying to solve the problem.

SUPER LOWER-THIRD TITLE:
Let the Customer Vent Frustrations

SUPER LOWER-THIRD TITLE:
Empathize and Show Sincere Interest

Mason is now finishing the spiel.

> MASON
> Now by my math, that comes to seventy-eight, fifteen. But, as always, your computer sees it as lower! Is that thing conspiring to send me to the poor house?

Wells smiles sympathetically.

---

page 6

> WELLS
> I'm sorry this whole thing has gotten you so upset, Mr. Mason. I

can understand exactly how you
must feel. I know how it is when
you depend on a place like our
bank to keep things in order and
they get (looks from side to side,
then whispers) . . . screwed up!

It's obvious this is having a
calming effect on Mason
SUPER LOWER-THIRD TITLE:
Focus on the Facts Not the Anger

HOST (VO)
And with Al on her side now,
Melissa moves to step three: focus
on the facts, not the anger.

MASON
It's frustrating, that's what it is.

WELLS
I know it is, and I want to make
sure, personally, that we get it
taken care of today, for good. Now
the first thing we need to do is get
at the facts.

Mason likes what he's hearing

page 7

MASON
Exactly. It's about time someone
showed the initiative to get to the
bottom of this.
(DIALOGUE UNDER)

Again, we stay on Mason. Wells
leaves the counter to get the
records. Al stews.

HOST
Melissa's sincere concern about
Al's problem and her plan for
getting to the facts have an
immediate calming effect on Al.
This is because, just like any irate
customer, what he really wants is
someone to take his side. In his
eyes, Melissa has done that. What
she's also doing is getting ready to
apply her final bit of strategy:
doing everything she can to try to
really solve Al's problem.

SUPER LOWER-THIRD TITLE:
Make Every Effort to Solve the
Problem

DISSOLVE TO

page 8

Several minutes later. Wells is just
returning to the counter. She has
Mason's statement and a
computer check sheet in her
hand.

WELLS
Thanks for your patience, Mr.
Mason. I know it took a while,

Mason can't quite believe this.

but it was worth it. I found the problem.

> MASON
> You're kidding!

> WELLS
> Nope. It's been right here under our noses all the time! I've got to apologize for whoever didn't clear this up from the very beginning.

> MASON
> That's okay. What was it? Was I right?

> WELLS
> Well, actually, yes and no. You see there was a charge you were leaving out.

> MASON
> What charge?

---

page 9

> WELLS
> A service charge for your credit card activity. See, it shows up (POINTS IT OUT) here. The problem is the way it's shown on the statement. It's really kind of hidden as a charge. So I consider it our fault, not yours.

> MASON
> So what do we do about it? Do I get a refund?

Wells smiles.

> WELLS
> I wish I <u>could</u> do that. I'll tell you what I can do, though. We have customer satisfaction meetings every Wednesday, and I'd like to present this to see if we can get this type of entry changed so it's easier to spot.

---

page 10

Mason isn't completely happy with the whole thing, but he does appreciate the concern and the fact that Wells is willing to take his problem to the "committee."

> MASON
> Sure. Well, okay, I guess I'll at least get some peace of mind with my statement from now on.

WELLS
That I promise. And, in fact, if you have any other problems, please let me know so we can get to the bottom of them right away.

MASON
Sure. Okay.

ANOTHER ANGLE
A much LONGER SHOT of the interaction. The Host now steps into the FRAME.

HOST
So there you have it. Four simple techniques that'll beat the satisfaction stealers every time.

DISSOLVE TO
Over FREEZE FRAMES of Melissa and Al's transaction, SUPER:

    Let the Customer Vent
    Empathize and Show Sincere
       Interest

    Focus on the Facts Not the
       Anger
    Try to Solve the Problem

HOST (VO)
Letting the customer get it off his or her chest. Empathizing and taking a sincere interest. Focusing on the <u>facts</u> not the anger. And last but not least, doing everything in your power to actually try to solve the problem.

page 11

DISSOLVE TO
HOST
She is beside the front door, just ready to leave.

HOST
And just who benefits from all this? Obviously the customer does, but you do too. Because you get to keep and continue to take home that job satisfaction that means so much. So really, <u>everyone</u> comes out the winner. <u>But why take my word for it?</u> Here are a few people speaking for themselves.

Host turns and exits.
DISSOLVE TO
INTERVIEW SEGMENTS
A series of brief interview bites from other employees confirming the value of beating the satisfaction stealers.

page 12

INTERVIEWEES
(points made in their own words)
• Professional customer handling <u>does</u> lead to more job satisfaction.

- Don't do it for the company, do it for <u>yourself.</u>
- Helps you go home feeling good at night.

DISSOLVE TO
EXT WELLS'S HOUSE—DAY
Wells is pulling in after a day's work.

          (MUSIC IN—UNDER HOST)
               HOST (VO)
Beating the satisfaction stealers. Sometimes it means counting to ten under your breath, <u>especially</u> when you know darn well the customer is wrong.

INT WELLS'S HOUSE
Melissa enters and greets her husband and child with a hug.

But in the long run it's worth it. Try it. See for yourself!

FREEZE FRAME
CLOSE-UP OF WELLS'S FACE.
CLOSING CREDITS

(MUSIC UP)

(MUSIC OUT)

FADE OUT

Commentary

As you read, "The Satisfaction Stealers" calls for a female on-camera host. The choice of a female was a conscious design factor based on the audience demographics of 80% female. The decision to have her appear on camera was based on audience analysis and budgetary factors. We felt that the basic philosophy of handling "satisfaction stealers" would be better accepted through personal interaction with a host who appeared to be a peer-level employee. In addition, the producer had the budget and client support to shoot in a local branch.

Note that the four parts of the first objective became the visual basis for the vignette segments. The continuing role-play visualized and dramatized the techniques through the actions of the angry customer and the experienced CSA. The superimposed support titles were used to reinforce the skills by appearing on the screen preceding each vignette segment.

The structure of this script is traditional and straightforward, based primarily on the client's and producer's request. It has a hook open in that it leaves an unanswered question: just what is this irate customer, Al Mason, about to do when he reaches the bank? The four instructional points form the body of the piece and are summarized in a brief close just before the testimonials. The testimonials provide a closing note of credibility on which to leave the audience.

There are several forms of visual transitions used in "The Satisfaction Stealers." Music, the opening title, and a dissolve take us from the introduction into the body of the program. A camera pan just prior to the actual vignette provides the transition away from the host. More dissolves are used, and freeze frames are the visual transitions into the closing titles summary. In addition, the host uses verbal transitions with lines like "let's rejoin Al Mason and watch an experienced CSA in action," "What she's also doing is getting ready to apply her final bit of strategy," and "here are a few people speaking for themselves."

Finally, the narration itself is light and conversational. The host doesn't seem to be stuffy or speaking down to the audience. Instead, her tone is warm and friendly, as if she were a peer or perhaps a supervisor. The narration also acts primarily as transition material, letting the vignettes—the *visual* elements of the script—tell as much of the story as possible.

18

□ □ □ □ □

Stock Footage Script

The scriptwriter's job would be ideal if the producers we worked for always had plenty of time and money. If this were the case, we could consistently write expensive, elaborate, perfectly visualized, and carefully executed scripts. Shortly after entering the scriptwriter's world, however, you will find that having "everything" is rarely the case. On almost every assignment, there is some compromise to be made. In most cases it centers around either time or money: a *lack* of one or the other. Perhaps an important executive meeting is being held in one week and your program is a session starter for the opening remarks. In such a case, there might not be time to send a crew on location to shoot the required footage. Or there may be plenty of *time* to shoot but inadequate funding.

I hope the majority of the scripts you write will be adequately funded, but you will surely run into situations where this is not the case. At times, there is a client need to be filled but almost no money to fill it. The producer is then left with a dilemma: how to get a show written and produced for virtually nothing.

One solution is to produce a stock footage program, which is made entirely from existing footage. The writer is usually given copies of the stock footage to review and asked to write voice-over narration to work with it.

"Centracall: The Power of Simplicity" is one such script. The client came to the producer on this project with very little money and the need for a program, fast. She was heading up a group who would be visiting business customer locations to sell a new service, Centracall. She had some fairly primitive print material, five employees, and a list of business addresses; that was it. She had been given the job, however, of making hers a profitable venture and was told she could have additional budget monies if and when she began to show some profit.

She felt a videotape would be an excellent sales tool to act as a meeting starter between her people and their potential Centracall customers. She was right.

The producer agreed. He was aware of two programs that had been produced previously on the same subject but from a different (employee information) slant. He thus had some existing footage on Centracall features sitting on the shelf. He arranged for the footage to be duped and called me in to write the script. After we had met and discussed the project, I immediately went home and spent several hours carefully viewing the footage and taking notes. I found that the producer's recollection of the shots had been right on. There was definitely enough footage to make a new program and, if it were scripted carefully, it could be used to represent an entirely new slant. The following is what we ended up with 2 days later.

CENTRACALL: THE POWER OF SIMPLICITY
A Two-Column Script
Developed from Stock Footage

FADE IN
CUSTOMERS—SOUND BITES
FROM SMALL-BUSINESS
OWNERS. STOCK FROM OPENING
SEGMENT OF A-9022.

CUSTOMERS
(points made in own words)
- CentraCall is simple to use.
- no equipment needed on premises
- contributes to bottom line by increasing efficiency
- can link several locations

FREEZE FRAME ON LAST
CUSTOMER—SUPER MAIN TITLE:

CentraCall
The Power of Simplicity

TITLE OUT
DVE TO
SMALL-BUSINESS OWNERS AT
WORK. STOCK FROM SECTION
TWO OF A-2234. AFTER GIRL.

(MUSIC UNDER THROUGHOUT)
NARRATOR (VO)
CentraCall. No doubt you've been hearing the word lately. What exactly does it mean? And more important what does it mean to you as a businessperson who depends on the telephone for your livelihood? In a word, CentraCall means power and simplicity.

FEATURES LIST—CG ROLL OVER
A POSTERIZED SHOT OF A
SINGLE-LINE PHONE
STOCK: A-4423, OPEN.

You see CentraCall is the power of some thirty plus state-of-the-art communications features, and it's the simplicity of a fingertip touch on a push-button phone. That's right. No equipment on your premises, no outlay for that equipment, no cables, connectors, closets full of hardware. Just convenience at the push of a button.

BUSINESS CUSTOMER USES FIVE
CENTRACALL FEATURES. STOCK
FROM A-4423. (FEATURES ARE
SUPERED)

Here are just a few examples. If you're expecting an important call but have to leave your desk for a meeting, call forwarding can divert your call to another phone.

NARRATOR (VO CONT)

Call <u>hold</u> then allows you to place that call on hold, if need be. Call <u>waiting</u> allows an incoming caller to receive a ringing tone, and it provides you with a call-waiting tone. If you need to, you can then transfer the call back to your office phone. At that point, you might want to <u>conference,</u> or share what's called a three-way call, with still another party. Simplicity and power. That's what CentraCall adds up to.

And just how does it work? Simple again. With the advent of computerized switching systems, we're now able to program in features that once required a good deal of electromechanical equipment.

SWITCHROOMS, ELECTRONIC EQUIPMENT, FUTURISTIC TECHNOLOGY. STOCK: A-2234. THROUGHOUT SHOW.

---

page 4

EARLY BUSINESS PHONE SYSTEMS IN U.S. STOCK FROM 0-1338 (FILM).

That not only means you don't need the equipment at your business, it also means you don't need a phone with multiple buttons, flashing lights, and so on. CentraCall is fully usable with any single-line push-button phone.

MODERN SMALL-BUSINESS PERSON ON THE PHONE. ("FOCUS" STANDARD OPEN)

So let's get back to the question, just what does all this mean to you as a businessperson? That we're not sure of. We're <u>very</u> sure, though, of what other people like yourself have been saying about CentraCall.

MORE CUSTOMER TESTIMONIALS FROM A-9022, NEAR CLOSE.

CUSTOMERS

- The greatest thing since push buttons.
- Don't know how I ever got along without it.
- Fantastic!

---

page 5

A-2234 SMALL-BUSINESS MONTAGE FROM OPEN OF SEGMENT THREE—IN MIDDLE OF SHOW.

NARRATOR (VO)

And that leaves just one final question: isn't it time you put the power of simplicity to work for you? If the answer to that question is yes, don't wait around. Give us a call. We're

NARRATOR (VO CONT)
Voice Link, and we're placing the
future on your doorstep.

VOICE-LINK LOGO—ANIMATED
VERSION.

(MUSIC STING, THEN OUT WITH
LOGO)

LOGO OUT
FADE OUT

Commentary

As you've read, the scene descriptions in this script are minimal, and
each has the library program number and general location of the footage
included. With this script in hand, the producer recorded a voice-over
sound track, found some upbeat music selections, included titles with
the help of a character generator, and made a short but very effective
program.

The client was extremely grateful and excited. The program helped
her venture get off to a great start. As a result, she returned for two more
programs later, both of which I was invited to script, this time for a bit
more money.

I've included this example for two reasons: to show you what can be
done with very little time and few resources and to make the point that
the scriptwriting process is sometimes subject to difficult limitations.

❑ ❑ ❑ ❑ ❑

PART IV

Putting It All Together

Revising Your Work 19

Client/Producer Revisions

Assume you've recently gotten your first scriptwriting job. You've worked diligently, incorporating all the writing skills we've been discussing. Your treatment and content outline have met with quick approvals. As a result, you've recently completed a first draft of the script. After submitting it to the producer, you've been waiting anxiously for several days. Finally, the phone rings. Sure enough, it is the producer. In a monotone she says, "I've looked it over myself and passed it on to the client. We both feel it's in the ballpark, but we've got some revisions. Can you get in for a meeting this afternoon?"

At this point, you probably start trying to read between the lines. What exactly is "in the ballpark"? Does it mean only a few minor revisions? How many and what types do they mean? Could it be word changes after all the time you spent on the dialogue? It couldn't be content inaccuracies; you know the subject like the back of your hand. Could they want major structural changes? Why is the meeting this afternoon?

After a few hours of nail biting, you make it in to the meeting and get the scoop. Things are not bad. There are three areas where the client feels the tone of the dialogue is too negative, a few minor content revisions, and a section the producer wants rewritten using a host and graphics instead of the role-play you decided on.

That's not bad at all, well, not *too* bad. Once you begin to really think about the changes, you decide they're not too *good*. In fact, you suddenly realize, they stink! Why? You disagree with them!

After all, you are the one who knows how to write scripts. You are the one who has gone through exhaustive hours of research, conceptualizing, writing, and editing. You are the one who has polished the work to a perfect focus, and you are the one who's been hired to communicate the message as effectively as possible.

After all this, you now become the one who is told by a client (a man who has spent virtually no time at all in the field with the people who actually do the work) that the dialogue in several scenes is too negative. The fact is you *know* it's negative. You wrote it that way because it truly reflects the temperament of the audience. It's negative because that is the best way to reach and identify with this audience. So you come to another conclusion: you are right and the client is wrong, no doubt about it.

Then there's the producer. Why should she want to give up a tightly written role-play scene that hits the nail on the head for a series of

Case Study 11   *The New Producer*

I was once hired to write a script for a large banking corporation. The subject was proper procedures for opening and closing branch offices. Opening and closing are the times when bank employees are most vulnerable to robberies, so you can understand how important the program was.

The project started smoothly. I met with the client and producer and worked out a verbal treatment for a dramatic program concept based on a robbery attempt foiled by two employees who had followed the proper procedures.

Following our first meeting, I immediately wrote the first draft of the script. The client and producer called me and expounded on the fact that they had been attempting to get a good script on this subject for a long time, and they now felt they had one that would show their employees how vital proper opening and closing procedures really were.

Because the project had gone so well, I was surprised when 2 weeks later I got a call from the producer inquiring if I would mind doing a major revision on the script. I assured him I would be glad to, but asked what problem had suddenly developed with the original script.

He assured me the original script had been fine but went on to say that a new producer had been brought in to do this project, and he wanted a "leaner" version of the script.

Soon after this conversation, I talked to the new producer. He wanted me to remove the on-camera host, do away with the dramatization, and eliminate the two other actors. It quickly became apparent to me that he wanted the script cut down to what essentially was a series of graphic screens and support text.

At that point, I told him that I would be glad to make whatever changes he required, but I wanted him to understand that I felt very strongly we would be taking all the impact out of a very important program and "carving" it down to a slide-show on tape.

He thanked me for my opinion and asked that I make the revisions as soon as possible. I did so and submitted the new version to him. He was pleased.

The point here is simple. Sometimes a script revision will be requested, even though you are *positive* it's not the right thing to do. At this point you must make a decision: do what you must and maintain a professional image, or refuse to make the changes and resign yourself to the fact that you may never be called by that company again.

---

graphics and a host talking head? The role-play works. It communicates. It's interesting. It's visual, which after all is the purpose of this whole exercise, right? No question about it. The role-play is the right way to get this information across.

The producer is wrong too. Neither she nor the client knows what they're talking about. They're asking for changes based on politics (the client wants everything to sound rosy even though it isn't) and personal preference (the producer probably wants to try out her new graphics system and maybe save some money).

So after all that work, you are now faced with the first of many similar dilemmas you will deal with as a scriptwriter. Do you go to battle for what you know to be right or be a creative wimp and change your material to make everyone happy?

This question is one that faces any writer — corporate or not — who accepts money for her work. You will face it too. Although the final answer, in your case, will be your own personal decision, one answer can be expressed in what I call the corporate scriptwriter's creed:

> As a corporate scriptwriter, I do the best work I can on every project, but I do not consider my scripts works of art. Rather, they are *art for work*. A producer is paying me to write a script that will meet *his* needs and the needs of his *client*, not my own.
>
>   When asked to revise a script in a way I do not like, I will present my case in a professional manner. If overruled, I will make whatever changes are required gracefully, professionally, and to the best of my ability.

In short, try to make them listen but never push too hard. Life isn't always fair, and neither is corporate scriptwriting. You should never consider damaging your reputation as a professional over a script.

You may well have hit the nail on the head when you said the client's changes are political and the producer's are personal or budgetary. You are also correct in assuming that these are not the best aesthetic reasons to change a well-written script. Nevertheless, they are facts of corporate life, and they often *are* legitimate concerns, whether aesthetically sound or not.

For instance, consider the client who wants the tone of the dialogue to be more positive. You may be right in assuming a negative tone will communicate better, but you may not be aware of a potentially bigger picture. Maybe negative approaches have been tried before. Maybe the client's vice president said specifically at the outset of this project, "I'll spend the money to do a video, but *only* if we can do something positive. I don't want to feed the negative attitude that exists in the field." Doesn't a positive tone then become a legitimate concern for the client?

The producer may want a graphics piece instead of role-play for any number of reasons. It could be strictly budgetary, always a legitimate concern, or she may indeed simply want to try out the new graphics system. Maybe this show will be her last chance to do that before making some major report on the value of the system to her boss, or she may simply feel graphics communicate better in that particular spot. Don't forget that she is the producer. If she has been developing programs for any length of time, she probably has an accurate feel for effective ways to present different types of material.

The critical point here is not these specific reasons. There are *countless* reasons for a client or producer to request changes in your

*Y*ou *have to find the thin line between fighting for your ideas and being willing to make changes. For the producer, it may be easier on the next show to call in a new writer than to have to wrestle a prima donna to get his changes made.*

— Alan C. Ross
Corporate Writer/Director

script. Some are fair and reasonable, and others are not. Some you will understand, and others will confuse you. As a corporate scriptwriter, you will have to be able to handle *both* on a continuing basis, with grace and professionalism.

Get Specifics

You should approach a revisions meeting the same way you would a content meeting. If you don't understand something, ask for clarification. This can sometimes pose problems because there are times when clients cannot easily state what it is they want changed. In these instances, if you don't understand what's in question, it is up to you and the producer to help lead her through the process of finding whatever it is. If it's left unresolved, you'll have a difficult time later trying to figure it out.

By following the creed, remaining professional, and being specific, you should be able to handle any changes required and remain on your producer's "good" list. With that position comes recognition as a professional who cares about her work but who realizes that at times change is required regardless of fairness or logic.

However, client and producer changes are not the only revisions you will make to your script.

Revisions by Committee

The dread of all writers is script revision by committee. You'll know you're in for it if, when you get in for that afternoon meeting the producer mentioned, there are *ten* people seated around a conference table instead of just him and the client.

In this case, each person will probably have a copy of your script covered in bold red marks. Each one will probably also have her own individual preferences about the script, and no one's preferences will be the same!

Unfortunately, revision by committee is sometimes a part of the scriptwriting process. Tact in dealing with it is very important. First and foremost, *never* get defensive. The worst thing any writer can do in a revision meeting is appear defensive about his work. This applies whether the comments made are valid or not. You should always be graceful and professional and convey the idea that every concern brought up is a valid one. Within these professional guidelines, however, you should also be ready to present your views as the expert on what works on the screen. You are, after all, the writer, and you have studied the subject in great depth.

Second, if a committee is to be the approving body, they should be prepared to reach a *consensus* on the final decisions. If revisions are left vague or unresolved, you should ask for clarification. It may take some time to get that consensus, but it will help you greatly when you're back home trying to incorporate the changes.

Your best bet with approval by committee is to try to avoid it in the beginning. This can often be accomplished by the producer at the outset of the project. He should recommend to the client that on this and all projects a single contact person be designated as the final decision maker.

---

*R*ead between the lines and try to get a feeling for what the client is trying to achieve. For instance, the client may have personal feelings that are different from what his boss has asked him to achieve. As a professional writer, you may be able to bring both perspectives together.

—Chris Van Buren
Corporate Video Client

---

*I*n script meetings, remember that the only way a writer can win an argument is by avoiding one.

—Daniel Gilbertson
Corporate Scriptwriter

In this way all committee recommendations can be considered and many weeded out well before your afternoon (perhaps also evening) meeting.

Director's Revisions

Besides clients, producers, and committees, probably the only other person who will have a major impact on your script is the director. As we have discussed, she is the person who has been given the ultimate responsibility of recording the script on film or videotape. She will be given your final version as a shooting script.

At this point, because all revisions have been made and the script is approved for production, the director's changes should primarily be centered around facilitating the production process. This usually means making changes in the left column or visual parts of the script to alter the way a scene would be recorded. For instance, she may see that a line of on-camera narration is too short to cover the visual scene description written in on the left side.

```
INT OFFICE—DAY
The host crosses the room, turns
on the computer, and shows us
the main program menu.
                                              HOST
                                Hello, I'm Jan Martin, and this is
                                Graphic Search.
```

This scene could play out as written, but it would probably leave quite a pause between the time the host introduces herself and when she gets the computer menu up on the screen. Because the director will actually be recording the material on tape, she must be sure that any such possible complications are taken care of before the shoot dates.

Therefore, she may want to rewrite the scene like this.

```
INT OFFICE—DAY
The host is seated at the computer
working. She realizes the
CAMERA is eavesdropping, looks
up, and shows us the main menu.
                                              HOST
                                Hello, I'm Jan Martin, and this is
                                Graphic Search.
```

With the host already seated at the computer, there will be no dead space or awkward pauses. Without her walk, the shot also becomes much simpler to light and record. These types of concerns are always on directors' minds, and this is another good reason for you to become as knowledgeable as possible about the production process. The more you know

*C*all the producer when your program is finished. Ask for a copy of the tape to see how the director visualized your script. You can learn a lot from it, and it shows the producer that you really care about the success of the project.

—Alan C. Ross
Corporate Writer/Director

about it, the more "shootable" your scripts will be. This saves the director and producer time and saves you the anguish of watching your script be revised still another time.

Personal Editing

One type of revision will have taken place long before your script ever reaches any of the people or groups we've been considering: your own personal editing. Personal editing is a way of life for any writer. It usually takes place as you are writing and often continues after you have set the manuscript aside for a cooling-off period until the time you finally give the script to the photocopier or put it in the mail.

For many writers, however, editing is not a labor. In fact, going through the editing process and watching a work come into clear focus can be just as exciting as the creative process of actually conceiving the material.

Writing is rewriting. Polish your work until every word counts.

—Daniel Gilbertson
Corporate Scriptwriter

Objectivity—The Key
The key factor in effective personal editing is *objectivity*. It can also be one of the hardest traits for beginning writers to develop. Perhaps this is because writing is such a personal and subjective form of self-expression. When first starting out, some writers feel that everything they put down on paper has some element of genius that shouldn't be tampered with.

As a writer gains experience, however, this perspective usually changes. Writing continues to be a personal process, but it becomes less like personal genius and more like personal *craft*. When writers enter this craft stage, most begin to see the rough edges in their work and become comfortable with true objectivity and personal editing.

If you are a new writer, you will probably need to pass through these stages in your own time. Perhaps your genius stage will be a short one, and you will be quick to see the values of editing; maybe your convictions are intense about your writing, and you will find it very hard to consider touching a word.

As a person who has been through both stages I can offer this advice: if you are not ready to edit your work objectively and allow outsiders to do the same, you are not ready to write professionally. This includes not only the corporate market but also just about any writer's market you can think of.

The Personal Editing Checklist
The basic elements of good writing apply to scriptwriting just as they do to print media. As a writer, you should be familiar with books like the Strunk and White's *Elements of Style* and know English usage, punctuation, spelling, grammar, and sentence structure.

Strictly in terms of scriptwriting, however, the following nine-point checklist should add to that basic editing knowledge. I would advise using it as a springboard for personally editing every script you write.

1. Is everything you've written simple and clear?
2. Will your scene descriptions create clear visual images for the director, producer, and client?

3. Is there a smooth sense of flow that carries the reader effortlessly through the piece?
4. Is your dialogue natural and credible, and does it make the point?
5. Is the narration smooth and conversational? Does it make the point?
6. Are you truly *comfortable* with the entire piece? If not, which parts make you uncomfortable and why?
7. Will the action you've written play out on the screen as well as it reads on paper?
8. Will the audience find your characters credible?
9. Does it accomplish the objectives you, the client, and the producer originally agreed on?

When you can comfortably answer yes to these questions, you're ready to send the script to the producer and wait for his call.

Finally

Revisions are often not the easiest part of the scriptwriting process. They are, however, an inherent part of every project and thus a process you must become totally comfortable with. The more projects you write, the closer you will come to reaching that total comfort level.

When you are able not only to write an excellent script but also work with clients, producers, and directors in accepting criticism gracefully and making revisions in a truly professional manner, you will be a valuable find in the corporate production world.

20 □□□□□ Getting Started

I remember as a young writer having the notion that one morning I would simply wake up and write something brilliant. Exactly how or why this sudden stroke of brilliance would occur I wasn't sure, but I felt it would, and from that day forward I would be considered an established writer. During that same period of time, I was hearing from writers and instructors, as well as reading in books and magazines, that the best way to become an established writer was simply by writing at every chance.

It took me a good deal of time and experience to finally draw an accurate relationship between these two statements. Eventually, however, I did, and it goes like this: good writing does *not* just happen one day, although many new writers seem to think so. People become good writers in one way only and that is by working at their craft.

I cannot overemphasize the accuracy and importance of this statement. Although writing is a creative skill, it is nonetheless a *craft*. Just like any other craft, it requires hours upon hours of practice to perfect.

Too many people make the mistake of confusing writing skills with "sensitivity" or "a creative gift." Although you may have a certain creative sensitivity or gift for various forms of expression, this does not mean you can simply sit down at a word processor and write well. A creative gift is only half the equation. The other half is gaining technical skill at whatever form of expression you choose.

If your form of expression is painting, you must learn hues, saturations, strokes, textures, colors, thinners, inks, and charcoals. If it is music, you must learn instruments, arrangements, scales, tempos, times, flats, and sharps. If it is writing, you must learn words, phrases, passives, actives, verbs, nouns, plots, characterization, structures, formats, dialogue, and narration.

In other words, a gift means you have the *potential* to be a great writer. What you actually make of that potential is a direct result of personal dedication, your willingness to put in the effort to perfect your craft. Any book leaves off at this point of personal dedication. So it is that this book will also leave off. In parting, let me leave you with a few final suggestions.

Read and Write

If you would like to write scripts, read and write scripts. If you can't land work, make up your own scripts. If you happen to see your neighbor fixing a tire, write a script on the steps involved in changing tires. Aim your script at a specific audience, like 16 year olds or unhandy executive types, and work out concepts, visualization, and the like. Go through the

From Intern to Assistant Director   Case Study 12

A young writer I know went from being a college intern to an assistant director in a very short period of time. She accomplished this through a combination of good work, good writing, and good timing.

She began as a "girl Friday" intern in the production facility of a large company. She let the people in the department know of her writing aspirations and over a period of months volunteered to do short, free writing jobs for a producer friend of mine. He liked her work ethic, her writing potential, and her eagerness to learn.

Several months later, he began producing a monthly company news program. Because the program was on a very limited budget, he could not afford to hire established writers. He sat down with the intern and made a deal: cheap scripts and hard work in exchange for steady writing jobs and lots of experience.

She began writing scripts for the news show at about half what the producer would have had to pay on the open market. This led to a natural evolution as a production assistant each month when the shows went into production. Production assistant eventually became assistant director. Her writing payments also went up, and all this led to a steady demand for her services.

Now the former intern is a valuable part of my friend's department. Her total elapsed time was 18 months.

---

whole process: needs analysis, content outline, treatment, and finally script development. In effect, place yourself in a write-for-hire situation, and follow it to completion. You may not gain the on-the-job experience you will eventually need, but this is an excellent way to get oriented to the types of documents you'll be asked to produce and what it takes to deliver them.

Reading scripts and other script-related documents such as treatments, needs assessments, and content outlines is another good exercise. You can find these in a number of places. One is in books. Go to the library or the local bookstore and check out what they have on scriptwriting. Another source is local college instructors, courses, and libraries. Still another way is to read up on the material used by local producers. You can sometimes do this simply by contacting them. Call and explain what your goal is, and ask if they would mind sharing copies of scripts and treatments that they have produced. Not only will this provide excellent reading material but also it may open doors to future relationships.

Check out the Business Environment

As we've discussed at several points in this book, the corporate scriptwriter should be at home in the business environment. This means knowing what businesses are in your area, checking out their buildings and offices, and learning something about their departments, structure, and people.

You can accomplish this by meeting with local producers or by contacting someone in the public affairs department of these companies or corporations. Again, be frank about what you're after and ask if there are press kits, organizational charts, company philosophy statements, and the like. Maybe there are even department heads or analysts who would be willing to meet with you and talk about their feelings on corporate video programs. If nothing else, you will gain a knowledge of the local business community and a certain comfort level simply by frequenting the business world.

The library is another source of business knowledge, as well as colleges. In fact, it wouldn't hurt to take a course or two in general areas of business operations or management.

Write on Speculation

When you feel you've gained enough skill to look for work, first try writing several scripts on speculation or "spec," as it's called.

You might simply pick a few "evergreen" business topics, such as customer handling, sales techniques, safe lifting, or safety in general. Write what you feel are effective scripts on these topics and send or hand-deliver them to producers. Tell them in your cover letter or conversation that you realize your approach is speculative, but your main intent is to break the ice and let them see what you can do (Figure 20.1). If they like a script, who knows, they may produce it or ask you to revise it or tailor it to their business. If not, at least you will have gotten your work read by a local producer who may remember the effort you put in when another project comes along.

Many a writer has gotten work not by seeing his spec work produced but by work offered to him as a result of a past spec presentation.

Internships

Some companies offer summer internships to college students on break. If you are financially able to survive on very little or no money, an intern experience will be invaluable, even as production assistant or helper rather than as a writer.

You can learn not only how scripts are handled but also the production process and the way the department works in general. Just knowing what a producer's day is like can help you learn to submit material that gives you the edge over other writers. Again, you will be developing and cementing personal acquaintances that should later be profitable business relationships.

Read *The Writer's Market*

Each year an updated version of this "spec writer's bible" is published. In it are sections on, among other topics, "Scriptwriting/Playwriting" and "Scriptwriting/Business and Educational Writing." Producers and distributors listed in these sections state the types of projects they are interested in reviewing and the payment arrangements they make. They often give

You have to sell your script. That includes explaining why you made the choices that you did.

—Alan C. Ross
Corporate Writer/Director

AUDIOVISUAL PRODUCTIONS

February 2. 1991

Martin Aspen
Employee Communications Manager
DARVEL MANUFACTURING
15144 Medford Forum
Danvers, MA 02188

Dear Mr. Aspen:

I've recently learned that Darvel Manufacturing is a long time
user of corporate video programing as a means of employee
communication.

I am a local college graduate with a strong academic
background in corporate scriptwriting, and I would like very
much to meet with you and discuss the possibility of writing
programs for Darvel.

Since I realize the words "college graduate" signal someone
new to the business, I felt the best way to introduce you to
my writing skills was simply to let them speak for
themselves. Toward that end, I have developed the short
enclosed script.

I arrived at the topic, warehouse safety, based on the
knowledge that Darvel does extensive storage and delivery of
heavy equipment parts. While I was not able to focus on
specific Darvel policies and procedures, the script is based
on careful research on the subject. And, of course,
customizing it for Darvel could easily be accomplished, should
you have an interest in producing it.

May I drop by and discuss this script and the possibility of
doing future work for Darvel? I will call in a few days to
introduce myself and find out if a visit is a good idea.

In the meantime, thank you very much for reviewing this letter
and my work. I look forward to talking with you soon.

Sincerely,

Bill Prentzer

Figure 20.1 An introductory
letter. It is important to have a
top-notch script sample to send
along with correspondence like
this. Without it, the letter may
go no further than the producer's
resume file or a waste basket.

tips to new writers on breaking the ice. A typical listing in *Writer's Market* looks like this.

> POINT PRODUCTIONS, 29401 Rock Tree Drive, Alondra Hills,
> CA 94201. Executive producer: Warren McKelvey. Produces
> material for business and educational markets. Buys 10–25
> scripts per year. Buys all rights. Enclose SASE. Fees: negotiable.
> Reports in 2 months.
> NEEDS: Employee sales training, management techniques and
> safety. Produces videotapes, slide programs, 16mm films and
> audiotapes.
> TIPS: We like to work with new writers. Do your homework.
> Think visually. Originality of message is a must.

You should buy a copy of *Writer's Market* and study it carefully. When you feel you've got something marketable, contact producers by mail or phone. If you don't make a sale, you'll at least gain a good deal of knowledge about the marketplace.

Business Cards and Accessories

When you do feel ready to move into the marketplace, you'll also need business cards. Letterhead stationery is not a must, but it's another accessory you might consider for sending those letters to producers and other businesspeople. Also get a leather briefcase and folder. You'll use these often when attending client meetings.

As we've already discussed, you should also dress and groom yourself professionally and carry a small portable tape recorder with plenty of paper, pens, and pencils.

The Computer: A Writer's Best Friend

If you're still using a typewriter, you should consider purchasing a computer as soon as finances will allow. A comparatively inexpensive home unit can save you hours of work and reams of paper.

IBM and IBM compatibles or "clones" seem to be the most popular. Apple systems, mostly Macintoshes, are also widely used, as are several other types. Word-processing programs on the market that are usable with these brands include WordStar, WordPerfect, MacWrite, Appleworks, Multimate, and others.

Cost and Components

The typical home computer costs between $1500 and $2500. Included in this package are the processor itself (the computer's "brain") with at least one disk drive and preferably an internal hard drive; a monitor or VDT (video display terminal), which is the TV screen on which you see your work in progress; a keyboard similar to a typewriter's; and a printer used to output your work onto paper.

How They Work

Word processing makes you much more productive as a writer because it puts a great deal of flexibility at your fingertips *before* your work ever gets onto paper. You can create, move, copy, format, change, delete, spell-check, and even grammar-check your work while it is still in "soft" form. Then, when you're satisfied it's ready to be seen, you simply print out the fully polished work and send it off. Storage space is also reduced because many of your stacks of paper will turn into backup floppy disks that are a fraction of the size and weight.

A Few Tips

I wouldn't profess to be a computer expert, but I've picked up a few helpful hints over the years.

1. Don't overpurchase. If your interest is strictly writing, put your money into an appropriate computer and good software. Don't get sidetracked into sophisticated graphics programs or other such nice-to-have features that won't contribute to your bottom-line income.
2. Purchase a good-quality printer. The printer's output is the *only* thing the producer sees. If it happens to be a poor-quality unit, no matter how much time you may have put in on the project, your work could *look* less than professional.
3. A computer with hard-drive storage will offer you ample room to run top-quality word-processing programs and also store your work. I find 30 megabytes more than adequate.
4. As you begin to work on a computer, *always* make backup copies of your work. They're quick and easy to produce, and they could save you hours, days, or even weeks of rework! Personally, I like to keep three copies of some projects: one on the hard drive, one on a floppy disk, and a third on paper.
5. In this age of technological breakthroughs, the benefits a computer can offer a writer are multiplying continually. The use of billing and money-management programs, modems, and on-line, exotic electronic services are some additional areas you may want to explore.

Finances

If you work as a free-lancer, you will need a method of invoicing your clients and a means of accounting for your income for tax purposes. Your invoice might be as simple as a letter to the producer stating the services you've rendered and the fees agreed upon. Figure 20.2 on page 178 is one example. Your accounting might be handled with a computer program or a simple, handwritten ledger documenting income and expenses, such as in Figure 20.3, page 179.

You should also discuss tax deductions with your accountant or tax preparer or read up on it yourself. As a free-lance writer, there are many deductions you may take that can help preserve as much of that hard-earned income as possible. The following lists some of the fairly standard ones.

materials	paper, pens, pencils, typewriter ribbons, word-processing computer disks, audio tapes, recorders, computer programs for word processing, computers used exclusively for word processing
travel	car mileage to and from writing assignments or writing-related activities
education	training to further yourself as a writer: seminar fees; course tuition, materials, books, and other related expenses

Alan C. Ross PRODUCTIONS
202 Culper Court
Hermosa Beach, CA 90254
Telephone: (213) 379-2015

INVOICE

                              December 15, 1990
                              Purchase Order #   D3657
                              Job: Turbo Fairing
                              Terms: COD

Judith Doud
V.P. Marketing
Zephyr Manufacturing
204 West Wind Blvd.
Los Angeles, CA 91301

Dear Judy:

Thanks very much for the opportunity to work with you on the
Turbo video. I hope it brings in more dealers than you can
handle. I'll give you a call in a couple of months to see how
the campaign is going.

                              Sincerely,

                              Alan C. Ross
                              President

acr:jar

Research and script writing, "Selling $1800.
Turbo Fairings"

       Less advance payments   900.

           Balance Due $ 900.

Figure 20.2 Sample invoice on letterhead paper. In this case, the writer wisely uses the opportunity to reinforce his relationship with the client or producer.

publications	books, magazines, or other publications you use to obtain work or stay abreast of developments in the writing field
postage	all postage for mailing letters, scripts, or other writing-related correspondence
photocopies	any copying expenses you incur in your work as a writer
phone	any calls made related to writing activities
accessories	any money spent on business cards, stationery, or other material used to market or advertise your services
entertainment	any money you spend entertaining producers or other people in the field as long as the entertainment is intended to produce or support your work

	DATE	DESCRIPTION	INCOME	EXPENSE
1	6/1	1st payment – Carlson script	500 –	
3	6/3	Ribbons – MAC		14.30
5	6/7	Xerox		6.12
7	6/9	Postage - spec script - "Safety…"		5.80
9	6/11	Rewrite fee - "Motivation…" workbook	215 –	
11	6/21	Lunch w/ Dale - Earthquake series		21.16
13	7/4	Phonecalls (8)		22.17
15	7/15	Paper		16.42
17	7/15	2nd payment - Carlson script (1st draft)	650 –	
19	7/29	Tape rental - 3 how-to's		9.45
21	8/4	ITVA dues		125.00
23	8/11	WRITER'S MARKET		19.95
25	8/13	Final payment - Carlson script	700 –	
27	8/21	Advance - Earthquake series (Hawfield)	950 –	
29	8/31	Mileage - 6/1 - 8/31 (342 × .21)		71.82
31			3075 –	312.19

Figure 20.3 Simple ledger sheet tracks income and expenses, in this case for 3 months. Additional documentation, such as receipts, is important as a supplement at tax time.

Fees

There are several options available for charging your clients. You can ask for a *per-page* fee if the producer queries you for a price; this is usually around $100 to $200. You might ask for a *per-running-minute* fee, which is close to the page fee because one page of script usually averages out to just under 1 minute of program running time on the screen. Another method is to ask for a percentage of the production budget, usually between 5% and 10%. Still another option, often the easiest, is simply to charge a *flat fee* based on what you feel the job is worth. As mentioned earlier, a typical 10- to 15-page corporate script pays about $1500.

Case Study 13   *Not Worth It?*

I know a writer who lost a good deal of work and cooled a professional relationship over a careless choice of words.

A producer called the writer to hire him for a quick rewrite. Because it was a relatively small job, the producer offered a payment of $500.

The writer happened to be very busy when he received the call. He declined the producer's offer, saying the project "isn't worth my time for only five hundred bucks." The producer resented the writer's seemingly aloof attitude, especially because he had given the writer a good deal of work in the past and happened to be under the gun on this particular project.

As a result, the producer stopped calling the writer. Several months later, the scripts the writer had been working on were completed. His work (and income) slowed down. He called the producer several times to offer his services, but no more work came his way.

The lesson here is one of common sense and tact. It's perfectly all right to turn down work if you have more profitable projects in the mill. There are ways to say no, however, that don't carry questionable undertones.

If circumstances such as extensive research or a particularly short deadline make the project abnormally difficult, the price can go up. It can also go up if your credentials are impressive enough. It can also go *down* depending on the company or producer you are working for, the budget available for the project, or any number of other factors.

In other words your fees are basically negotiable. Where you should start is, of course, up to you, but my suggestion would be to supplement your income with other work in the beginning and start your writing career by working cheap, maybe even free. Corporate belt-tightening and ongoing budget crunches are problems for producers, and a very attractive rate or an introductory freebie could encourage a producer who might not otherwise use you to give you a try, especially if she happens to have just inherited a low-budget project. Once you've broken the ice and shown what you can do, you can then move up to the going rates.

*W*rite for the long haul. Every script you write is an opportunity to keep a client and create new business.

—Daniel Gilbertson
Corporate Scriptwriter

You should receive your payments promptly, either in milestones as the project progresses or in one payment upon script approval. If you don't get paid, you should contact the producer and tactfully request an explanation. If he then makes payment, you're set. If not, you should refrain from working for him again until paid for the previous job. Even then you might ask yourself if you care to continually do battle for your money with that producer. It may be better to look for more stable companies to work for.

In Closing

For those of us who write, the subject is an inexhaustible one. There is much to learn, much to do, and much to think and talk about. At some point, however, all this becomes inadequate. It is time to write. I hope you have now reached that point, and I hope this book has helped you arrive there.

I also hope that the initial question posed in the very first pages of this book, What exactly is the right way to write a script? has now been answered in your mind. The answer, of course, is not according to any single book, person, opinion, or format, but rather in whatever way, in keeping with industry standards, will be most effective at accomplishing the goals of each individual project. That goes for the needs analysis, content outlines, treatments, concept thinking, visualizing, and every other aspect of scriptwriting we've discussed. And with that . . .

INT DEN—NIGHT
Back to the writer. He quickly types a few final words into the keyboard of his computer. He checks the screen. The words are right. He removes his glasses, rubs his eyes, and smiles. As he gets to his feet and exits the SHOT, we begin a SLOW DOLLY IN to the computer screen. At the same time, HEAR him say . . .

> WRITER (OS)
> Sweetheart?

> WIFE (OS)
> Out here, hon.

> WRITER
> Guess what?

> WIFE (OS)
> You're done!

> WRITER (OS)
> Finally.

> WIFE (OS)
> Okay, I've got to know. How did you
> end it?

> WRITER (OS)
> (after a pause)
> What would you say if I told you it was
> going to sell a million copies?

DOLLY ENDS. We are now in the EXTREME CLOSE-UP. The last words on the screen are: "Good luck!"

> WIFE (OS)
> (with a chuckle)
> Ha! I'd say, "Good luck!"

                              FADE OUT
> THE END

Scriptwriting Steps

Because our discussion of the steps typically involved in a scriptwriting project have been spread over many pages in this book, the following quick-reference overview of the process might come in handy. Keep in mind they are offered as *general guidelines*, and different producers and companies will have their own ways of handling scriptwriting projects.

Scriptwriting Project: Typical Steps

1. A client calls or meets with a producer. She requests a video program to solve a communication problem. The producer verifies that a program is the right tool for the job. (Verification may involve the development of a program needs analysis by the producer.)

2. The writer and producer meet in person or by phone. The producer hires the writer. A discussion of fees, parameters, time lines, and deliverables takes place.

3. An initial client-producer-writer meeting is held. The primary discussion focuses on objectives, purpose, audience, utilization, cost, and deadline.

4. If not already prepared, a program needs analysis may now be developed by the producer or writer. This provides a basic analysis of the instructional design elements important as a foundation for the program. It is sent to the client.

5. The client approves the program needs analysis. (The program needs analysis may *not* be requested by the client or producer. In that case, the writer should still do design research.)

6. The writer does content research, which includes calls, interviews, gathering and reading written material, and attending meetings or classes. (In some cases, a content outline is developed at this point. If so, it is submitted just as all other documents are for producer and client approval.)

7. The writer conceptualizes, visualizes, and develops a treatment. The treatment describes the program visually in simple, narrative form.

8. The treatment is submitted to the producer. If the producer approves, he submits it to the client. If not, the producer meets with the writer and requests revisions. The producer then sends the revised treatment to the client.

9. The client reviews the treatment and approves it or suggests changes. Content experts may also review the treatment. (If changes are minor, development may move into the script

stage. If changes are major, a new version of the treatment should be developed.)

10. The treatment is approved.

11. The writer develops a first draft of the script based on the approved treatment. He submits it to the producer.

12. The producer reviews the first draft. If he feels it is ready for client review, he submits it to the client. If not, the producer and writer meet, and the writer revises per the producer's request.

13. The producer sends the first draft to the client.

14. The client reviews the first draft. She also shares it with content experts if appropriate. The client and content experts approve the script or return input to producer and writer.

15. The writer develops and submits a second draft based on input received on the first draft.

16. The second draft is submitted to the producer. The producer again submits it to the client.

17. The client and content experts provide approval or further input on the second draft. The draft is returned to the writer and producer.

18. The writer develops and submits the final draft.

19. The client and producer approve the draft as a shooting script.

20. Preproduction begins. (Possible on-the-spot revisions or rewrites may be requested by the client, producer, or director during preproduction.)

Two Business Proposals

No doubt you will someday find yourself in the position of having to write a proposal for a prospective client. Although effective proposal writing is an art in itself, the two examples that follow should at least give you a starting point on which to base your own.

The first was written for an executive audience. The objective was to give the video department in a national corporation a "foot in the boardroom door" during a time when money was becoming extremely tight.

As you might guess by reading the proposal, a key buzz phrase in the company at this time was "added value." The recipients of that added value were to be the company's customers and employees. In fact, the company had just kicked off a major internal campaign called "The Value Machine."

The second proposal was developed for a much smaller company. It came about as a result of an informal introduction by a friend and a brief subsequent meeting. In this case, the president (and owner) of the company was just getting started with a half-dozen frozen yogurt stores. He needed a series of print and video training programs to educate his employees on what he felt were the keys to nurturing a successful business.

Just as with the scripts and other writing samples in this book, as you read these proposals, consider what you feel are their strong and weak points. Also ask yourself how you would write such a proposal. In fact, why not do research on a local business and actually write one? What better way could there be to get that first experience and, who knows, maybe that first job?

---

page 1

A NEW EQUATION FOR VISUAL MEDIA IN THE 90s
by
Ray DiZazzo

OVERVIEW

This document proposes an innovative new way for visual media to assist L&L/P–West Area in achieving its 1991 and 1992 goals and objectives.

THE TRADITIONAL ROLE OF VISUAL MEDIA

Visual media (training, motivational, and informational videotapes) have played a key role in L&L/P's evolution over the past 15 years. Literally hundreds of programs have been produced, saving countless labor hours, preventing costly accidents, and helping to avoid many other problematic situations.

Until now, the work of the media group (L&L/P ComVid) has been carried out on an "as needed" basis. Programs have been requested by staff and field employees as various communication problems have arisen.

VISUAL MEDIA IN THE 90s: A NEW ROLE

As the industry and our company change radically, we at ComVid feel that now is the time to change the way our services are traditionally used.

We see a need to integrate our work into the most basic goals and objectives of L&L/P–West Area. By carefully aligning our visual products with the basic business of planning and achieving the company's continued growth, we are confident we will be able to focus our efforts more efficiently on helping to assure a profitable and healthy future.

Toward this end, we have carefully reviewed the "L&L/P 1991, 1992 Operating Plan" and attended the recent West Area Key Managers Meeting.

Using these two items as a "creative foundation," we have conceptualized a number of potential visual media projects and the benefits to be gained from their development.

page 2

PROGRAMS AND RATIONALE

Value has become a key word in our vocabularies, and it appears to be one that will be with us for some time, perhaps years. Two types of value were recently cited by company president Anthony Hall as most important to our future: value to our customers, and value to our employees. With these two concepts in mind, we propose the following:

Executive Cable Talk Show: "Community Connection"

A series of executive or management talk shows, produced at the L&L/P ComVid for broadcast on local cable channels, would feature key L&L/P managers discussing issues of importance in both the industry and the community.

For instance, the safety manager or director might be the interviewee on a program dealing with 911 and/or emergency evacuation techniques. Stock footage, shot for use in previous internal L&L/P programs, could be used in a program such as this.

The subject of long-distance companies and their place in the telecommunications industry (always a confusing subject) might also be covered, and, of course, there would be segments to highlight some of the typically "invisible" services we provide, such as 4-tel and TAC centers.

Another of the many possible "Community Connections" might include the company doctor discussing health topics and (using our fitness center as a setting) possible fitness techniques.

Yet another segment might feature an installation representative discussing typical phone problems that can be solved in the home.

The underlying message contained in this series comes down to that familiar word, value, the added value of a company that cares enough about its customers to air a series of television programs on their behalf.

NOTE: AIR TIME ON COMMUNITY ACCESS TELEVISION IS FREE, AND THESE PROGRAMS COULD BE "BUNDLED," THAT IS, PRODUCED IN QUANTITY AT A SUBSTANTIAL SAVINGS.

Continuing to focus on the words <u>value</u> and <u>customer</u>, the following would also be a unique offering:

Customer Take-Home Programs

A series of customer take-home videotapes designed to explain our products and services and pass along handy telecommunications hints to the general public would be given to customers (perhaps at Commumarts or business offices or sent by mail) as an added value received only when they do business with L&L/P–West Area.

The programs would consist primarily of "tabletop" product shots of our telephones and equipment and stock footage already used in the company for previous internal projects. In addition, the programs would utilize attractive on-screen graphics, pleasant music, and voice-over narration explaining all facets of our service and equipment.

Produced in this way ("bundled" and utilizing mostly stock footage), they would be very inexpensive.

As Mr. Hall has said, however, value is not something we should offer <u>only</u> to our customers. Employees, too, are a critical part of the business equation. With this thought in mind, we propose

Employee Take-Home Programming

A series of employee take-home videotapes on the subjects of personal and professional growth would cover such topics as

> working toward management
> the telecommunications business and you: a look at tomorrow
> job and family: a rewarding balance

The programs would offer our employees the <u>added value</u> that confirms "L&L/P truly does care about you and your family, and we want to help make your future a positive, productive one, both on and off the job."

Another word gaining daily importance in our business is <u>communications</u>. As one of the key focus points brought out in the employee opinion survey, it has become one benchmark by which we will measure our future success.

As we all know, one of the key persons in the employee communication link is the first-line supervisor. Through our supervisors, L&L/P's thousands of craft employees gain the majority of their information and maintain their connection to the world of management.

In order to help assure this vital communication link gains effectiveness as the months progress, we propose the following.

"On the Line"—Supervisory Development Series

A series of programs aimed specifically at L&L/P first-line supervisors, carrying the main title "On The Line," would carry subtitles (and thus cover topics) such as

"Gaining Employee Respect"
"Coach Management"
"Principles of Motivation"
"You Are the Critical Link"

These programs would accomplish two important goals. First, they would train first-line supervisors in the effective techniques of supervision needed to keep L&L/P's "Value Machine" running at full speed on a daily basis. Second, they would provide that employee with added value by letting first-line supervisors know that L&L/P really does understand their importance and is supportive of the sacrifices they make.

As a custom-made series, these programs would be L&L/P specific. They would focus not on general supervisory issues but on our own problems, situations, programs, and evolution as a basis for supervisory development and change.

page 5

COST AND DEVELOPMENT METHODS

Although video programs obviously cost money, the programs outlined in this proposal would be very inexpensive as such projects go. We estimate them to cost approximately one-third the going rate for such programs on the open market.

The three keys to this reduced cost are

> bundling, that is, producing multiple programs at once

> the use of stock footage, previously shot or internal use

> the stringent cost and quality controls exercised on every project by the ComVid staff

SUMMARY: THE NEXT STEP

We at L&L/P ComVid see a profitable and successful future for L&L/P–West Area. We hope to share in the challenges of helping to create that success. We feel confident that the programs and methods outlined in this proposal will be a valuable asset in accomplishing that end.

If the L&L/P–West Area executive team shares our excitement for what we've proposed, we would welcome the opportunity to take the next step. We would like to develop a series of program outlines and associated budgets that would reflect specific program content and visualization as well as accurate time lines and cost estimates.

We look forward to your response and the opportunity to add ComVid value to the L&L/P–West Area "Value Machine."

page 6

VIDEOTAPE & WORKBOOK TRAINING SERIES
A Development Proposal
For the Parfin Group and Arctic Place Yogurt Stores
by
Ray DiZazzo

OVERVIEW

This proposal outlines the development of a series of five videotape programs and supporting print material. This material will be used by the

Parfin Group to train and motivate new Arctic Place employees. Although additional facts are needed to finalize program content and develop specific training objectives, a basic structure for the series, general program concepts, and rationale are provided on the following pages.

<u>PROGRAMS</u>

In keeping with the spirit of Parfin's existing policies, these programs will be generally light in tone and fun to watch. At the same time, however, specific training objectives will be established for each program, as well as a means of evaluating their effectiveness. In short, employees will enjoy the programs but come away from each viewing with knowledge that will assist them in moving into their jobs comfortably and efficiently.

With these facts in mind, a suggested series title for the programs is "Arctic Previews." Each program would also have a subtitle descriptive of the subject being taught.

The suggested program titles and a brief synopsis of each are

1. Arctic Previews One: Cool People
   An introductory overview of the four main "ingredients" most important to "cool" people (customers): a quality product; a neat, clean personal appearance; a positive attitude; and store cleanliness. Customer-on-the street interviews will be included in this program and a brief welcome by company president Bill Parfin.

2. Arctic Previews Two: Counter Courtesy
   A program using short vignettes to stress the importance of handling customers in a friendly, helpful, and courteous manner. Vignettes will focus on the right way to handle customers and the resulting self-satisfaction. Personal appearance will also be a part of this program.

3. Arctic Previews Three: Exquisite Chill
   A how-to program showing the approved, step-by-step method of product preparation, display, and presentation to customers. Employee vignettes will provide visual examples.

4. Arctic Preview Four: A Clean Sweep
   This program, again using on-the-job vignettes, will cover the basics of maintaining a clean, healthy work environment that is attractive and comfortable to Arctic customers.

5. Arctic Previews Five: Cashing In
   Another step-by-step, how-to program, this time on cash register operation.

<u>SUGGESTED FORMAT</u>   *page 7*

A standard open and close should be used on every program. Each would begin with a brief series of unscripted testimonials made by veteran Arctic Place employees. These would stress the importance of doing a good job and the reward and self-satisfaction that results.

These would be followed by a fast-paced, upbeat montage highlighting general work activity both up front and behind the scenes at Arctic Place stores. These shots would be supported by pleasant music and might include close-ups such as a cup of yogurt being decorated with colored sprinkles, an employee smiling and conversing with a customer, a child eating yogurt and obviously enjoying it all over her face, a cash register ringing up, and ingredients being mixed.

The last shot of the montage would freeze, and the main title would be superimposed. Beneath it, the subtitle would then appear. Finally, this shot would dissolve away, leaving us inside a typical Arctic Place. It is here we would find our host, perhaps just finishing a yogurt, chatting with an employee, or enjoying a newspaper at a corner table.

The host would look up, realize we've "arrived," and welcome us. He would then introduce us to the subject of the program. From this point on, he would act as a kind of guide, reappearing from time to time to make important points or to help us make a transition to a different topic. His manner would be light, conversational, and friendly, and his dress would be informal.

Between host segments, employee vignettes would demonstrate the skills being taught. Additional employee testimonials would also be used to support each message. Superimposed titles would reinforce key points.

Just prior to the close, the program would return to the host for a brief wrapup. He would then bid us farewell and leave for another Arctic Place from which to bring us our next program.

At the close of each program, credits would roll, identifying and thanking employees who have contributed, either through on-camera interviews, as vignette actors, or by providing behind-the-scenes support and coordination.

page 8

RATIONALE

The format described would be effective for several reasons. First, the standard open and close would give the series a positive sense of continuity and personal "signature." Second, the heavy use of employee testimonials would result in a high level of program credibility. Aside from music videos, beach reports, who's going with whom, and the latest fashion fads, the young adults are most likely to give their full attention and trust to people their own age. Third, the same familiar host also becomes a figure of authority and trust. Finally, the use of employees instead of professional actors in vignettes would lower the overall budgets.

SUPPORT WRITTEN MATERIAL

"Arctic Axioms: In Black and White" would be the title of a stylized three-ring binder used to support the video programs. The binder would be designed in glossy black and white with the Arctic logo on its cover or perhaps a stylized tuxedo or penguin design. It would contain written material developed to provide additional detail on the principles covered in each program.

A copy of the "Axioms" would remain at each store and be updated if additional programs were developed. The facts would act as a handy reference for use at any time by new employees or perhaps as suggested reading by local managers.

The "Axioms" would also reinforce the personal signature of Arctic training and be in keeping with the fun tone of the overall training method.

COST AND TIME LINE

Assuming the five programs are developed concurrently as outlined in this proposal, the entire series and related print material would cost approximately $81,000.

Development would take approximately 7 months, as follows

script research and development	3 months
preproduction (planning and scheduling)	3 weeks
production (recording all material)	2 weeks
postproduction (editing)	3 months

PAYMENT

page 9

One common method of payment on such a project is in milestones. These could be divided, for the most part equally, and made on agreed-upon dates coinciding with project start-up, treatment approval (all shows), script approval (all shows), production start-up, rough edit approval, and program delivery.

SUMMARY

The programs and print material outlined in this proposal offer a unique and effective method of training new Arctic Place employees. These programs also offer a wise investment in the future because they will allow Parfin management to send the same positive and valuable message to new employees for years to come.

Following a decision to move forward with the project, development could begin at once. As a writer and producer of training materials, I would be both proud and excited to be a part of that development.

"You Script It" Answers

You Script It 1

In a very professional and tactful way, you should say something like this (of course, not all in one big chunk): "I'm sorry to hear you're going to be so busy. I have a few reservations, though, about just moving ahead with none of your input at all. I'm sure we can do the job just fine, but client support in the form of input and opinions is usually a real necessity. Is there someone you might delegate to look over the material we develop, perhaps a content expert or colleague? Maybe we or that person could also fax you the material, or we could call or send it by mail. It also occurs to me your boss and the committee members should follow along with the development. That way we can be sure the project is in tune with their expectations also. I'd hate to get all the way through several weeks of research and script development and go in a direction you or they aren't expecting."

What you've done with this little monologue is express some legitimate and very important concerns. Uninformed bosses, absent clients, and committees left in the dark are all analogous to the words "major rewrite."

Your producer should know this. If you've made your points tactfully, she should back you up at this point. If she doesn't and both she and the client insist on moving forward, you probably have no choice. You should then proceed on the most accurate information you have and hope for the best. Later, if the project does run into trouble, at least you will have spoken up about your concerns.

Your other option would be to try to decline gracefully on the project, but I wouldn't recommend this. Pulling out of a project at a time like this would probably be viewed as "desertion" by the producer.

You Script It 2

This is a case of misinterpretation due to a vague objective. "To assure that employees can use the Excet computer system to budget and word process" leaves far too much to interpretation.

In this case, the client has interpreted it to mean his employees would be *totally functional* on the Excet computer. This is too much to ask of a videotape program, even a good one. Your interpretation and probably the producer's was that "able to use" meant *partially* functional.

Writing that first objective like this would probably have saved the misinterpretation:

1. Having viewed the proposed program, audience members will be able to
   a. turn on the computer
   b. create a basic file using the word processor program
   c. create a basic budget sheet using the budget program

You Script It 3

This is a tough one. To go straight to program concepts without considering the design elements your client has referred to as "school-type stuff" is really the wrong approach. Based on what she's saying, however, telling her this may not be your best bet at the moment.

You would probably be better off putting the objectives aside *for a while* and moving on to the subject of concepts. This would allow her to get out some of the ideas she's obviously dying to talk about. These should be taken with a grain of salt, and the design information you've started to acquire is still necessary.

You might consider asking the client if you should get it from someone she delegates, a peer or subordinate. You might also try telling her something like, "I know things like objectives and audience analysis sound pretty boring, but they only act as a *foundation*. I'll go on to develop some exciting and innovative ways to get your message across. I just want to be sure they're the right concepts for your situation. Strange as it may sound, it's things like objectives and audience analysis that help me do that."

You Script It 4

Your best program concept in this case would probably be the documentary. When employees are suspicious about a company program, they are most likely to believe other employees. Footage of those other employees voicing their initial fears and current satisfaction with the new system would be extremely credible to this audience. It would be very important, however, that they are *unscripted* interview pieces spoken in the employees' own words. Putting scripted words in their mouths would destroy the documentary flavor and the credibility of the program.

If the program objectives were only to *demonstrate* the system and if the audience analysis hadn't brought to light the mistrust and skepticism present, the dramatic approach would probably work well. In this case, however, it would do little to alleviate employee fears. In fact, it might heighten the anxieties by making the employees feel they are being conned by management. At best, audience members would probably feel a bit insulted, sensing that the company was trying to quell their fears by creating a "convenient" little story with a happy ending.

The host on camera is a better concept than the dramatic approach. If he were a believable type and the script not condescending, at least it would be a straightforward presentation instead of a "story." Even if the host were very good, however, and the intercut footage of the other employees showed a wonderful system, the credibility and sense of reality created by the documentary probably could not be matched with this concept either.

You Script It 5

An audience of "good old boys" would probably get a bellyful of laughs out of the role-play idea. It would become a situation in which a program attempts to convince people to change their attitudes by presenting a story that conveniently has all the proper actions and perfect answers.

The music video would no doubt do the same thing. Somehow I can't quite imagine a group of tough-as-nails construction workers tuning in to a music video on safety. If it were presented as a farce, though, they might just watch it and even tune in. The question then becomes whether it would have any chance at *changing their behavior* and making them more safety conscious. The answer is probably no.

This same question applies to the comedy idea. They might watch and enjoy it, but would it make them think seriously about safety? Would a series of comical vignettes make them remember the safety procedures and follow them? Although presenting material in a unique way definitely makes it more memorable, in this case memorable might not change their minds and make them really act safer.

The final concept, however, probably would. In this case, the host figure would be somewhat of a peer. If, as the synopsis states, his attitude is positive but unsympathetic, he would have a good chance of getting through to this audience.

Another positive element in this concept is the idea of letting the employees themselves make the decision. There is something much more attractive about a new idea presented in a "you decide" fashion instead of "do it whether you like it or not." This probably holds true for most of us, but for this audience I think the independent decision factor is even more attractive.

Finally, this concept incorporates the *family* element, another factor dear to the hearts of all of us but maybe more so to this audience.

What all these elements add up to is perhaps not the most entertaining of the concepts presented, but then motivation to change attitudes is the objective, not pure entertainment. With that in mind, this last concept is probably the most likely to pay back its production expenses in the form of lower accident costs to the company.

You Script It 6

This treatment segment is much too cluttered with clumsy technical script terminology to allow the reader to concentrate on the story itself. At best, it is choppy, disjointed, and seemingly too complicated. It would have been much better written like this.

> Mary and John walk into John's office discussing Arnold's poor performance ratings. Arnold passes by, hears what they are saying, and comes in. He is totally distraught.

Two days later Arnold and his wife, Jennifer, discuss the rating at dinner. Arnold says that because of it he is thinking of quitting.

You Script It 7

Although it's not really fair of this manager to put you on the hot seat with his request for an on-the-spot concept, sometimes this happens. To tell him you can't do what he's asking wouldn't be the best answer in this case. He's obviously on the spot himself, and he's looking to you for an out. If you could give him that out, you'd probably be laying the foundation for a long and profitable relationship.

What you *should* do, then, is shoot from the hip but make him aware that that's exactly what you are doing. Then, quickly consider what you know about your audience, and those three key words: *tough, competition,* and *survival.*

One large company I know of handled such a need with the use of a sports analogy. They hired a well-known athlete as a spokesperson and developed a script that dealt with the need for urgency and toughness in competitive situations. A parallel was drawn, of course, between the company's tough market position and the athlete's competitive sport.

Stock footage of the athlete in action came as part of the package when he was hired. This was mixed with footage of employees at work and employee interviews on the subject of business competition. The result was an effective, attractively produced, well-received program.

You Script It 8

There are two primary faults with this very awkward bit of dialogue. The first is the unnatural speech patterns. Rather than letting the conversation move back and forth as it normally would, the writer has clustered several ideas into a single paragraph for each speaker. The second major problem here is a lack of contractions and colloquialisms. Just like the idea clusters, this results in very *un*natural sounding dialogue.

Both these faults in the same piece add up to a horrendous piece of writing. Your more natural version of this conversation should read something like this.

JEFF
I've seen how you handle the
paperwork lots of times.

MARY
Good. What've you figured out?

JEFF
You do the math first.

>   MARY
>   (coaxing)
>   Right. And?
>
>   JEFF
>   And you type the envelopes second.
>
>   MARY
>   Right again. So you want to give it a
>   shot?
>
>   JEFF
>   I'd love to!
>
>   MARY
>   All yours.
>
>   JEFF
>   Great! And I promise I'll do a good job.
>
>   MARY
>   I'm not worried at all. I'm sure you'll
>   do fine.

You Script It 9

Making a host and titles more a part of a program's "show" elements involves giving them characteristics that increase their visual interest for the viewer. For instance, if the host appeared in a location visually relevant to the program, his segments would have increased visual impact.

As an example, suppose the first safety rule involved driving defensively and the host appeared at a wrecking yard where a company truck totaled in a recent accident had been towed. This would definitely add to the visual quality of his role.

As for the titles, if they were cleverly animated as they appeared in the program (something easily done with most character generators) and superimposed over, say, dramatic freeze frames from each of the vignettes, they too would be visually more involving to the viewer.

You Script It 10

A miner who, in his greed, digs himself into a hole too deep to get out of might work well as a program concept for dealing with dishonest behavior. This kind of visual analogy could easily draw a parallel to a person who, say, continues to cheat or lie on the job. It could even be handled with an element of humor to lighten the impact of discussing such a delicate topic.

A person unknowingly painting herself into a corner in a vacant room could also work, as could a gravedigger who finds out the grave he's been digging is *his own*. Still another analogy is that of a gambler who in the end loses all.

Glossary

ambience The background noise present at any location. Room ambience is also called room *tone*. Ambience on a street corner would include traffic sounds such as horns honking and perhaps pedestrian chatter.

angle on A camera term noting that the angle should be on a person or thing but not specifying a focal length such as MEDIUM CLOSE-UP or CLOSE-UP.

audio The sound portion of a script: narration, dialogue, sound effects, and music. In the column format, the sound or audio column is usually written on the right side of the page. In the screenplay format, it appears in both the scene descriptions and in the dialogue or narration column down the center of the page. *Audio* also refers to the sound portion of a tape or film recording.

audiovisual (A/V) Any type of sound (audio) and picture (visual) presentation. A/V is often used to describe a slide or overhead presentation instead of a film or videotape production.

BG A script abbreviation meaning background.

big shot A very wide establishing shot.

camera A script term usually used to explain what the camera is *doing*, for example, CAMERA DOLLIES LEFT past the desk REVEALING John.

client The person requesting, approving, and often paying for the program. As such, the client is one of the writer's key business contacts. The other key contact is the producer.

close-up (CU) A shot of an actor or item that fills the majority of the screen. In a CLOSE-UP of a female character, for instance, we would see her entire face and the upper part of her shoulders. In a CLOSE-UP of a glass, we would see the entire glass and a small area surrounding it.

column format Script format in which the sound and picture elements are separated into two columns on the page. Usually the picture column is on the left, and sound is on the right.

concept The basic premise upon which the script, or segments of it, will be based. Concepts link content and design with visualization and story line.

content expert A person who is an authority or expert on the subject the writer is researching. A writer will often be assigned a content expert by the client or producer while developing a project. Also called an SME, meaning subject matter expert.

content outline   A factual outline created by the writer. A content outline organizes facts into a properly structured format from which to
develop a treatment and script. Often sent to the producer and client
for approval, a well-written content outline is also an indication
that the writer is on target with his assessment of the proper facts to
be included or deleted.

corporate culture   The basic philosophy, policy, and standards systems a
large company or corporation uses as a framework for its day-to-day
business activities.

corporate television   Television productions produced for an in-house
audience of company employees. Also called *industrials* or *business
productions.*

cut   An instantaneous picture change used as a transition between
scenes; also used *within* scenes to describe viewing the action from
different angles.

dialogue   Conversation spoken between at least two actors in a role-play
situation.

director   The person responsible to the producer to interpret and execute
the written script into film, audio, or video footage appropriate for
editing into a completed program.

dissolve   A transitional effect in which one picture fades out as another
fades in. Most often it is used to suggest a passage of time or major
location change. A dissolve can also be used to "soften" jarring or
abrupt cuts.

dolly   Used as a verb, a forward or backward movement of the camera
during a shot. This is opposed to a TRUCK, which is a left or right
sideways camera motion. TRUCK and DOLLY are often used synonymously. Used as a noun, a unit with wheels on which the
camera can be mounted.

draft(s)   Versions of the script. Typically a scriptwriting project goes
through about three drafts. The first draft is the writer's initial
version. This is given to the producer and sent to the client. Revisions are then worked into the second draft. The third draft is often
considered the final draft or shooting script.

DVE   Abbreviation for digital video effects that is often used in scripts
to precede a description of a digital effect, for example, DVE—
PICTURE FLIES OFF SCREEN LEFT.

entrance   An actor's cue to move onstage or into the camera's frame.

establishing shot   Usually the first WIDE SHOT of an environment.
Sometimes it is written WIDE SHOT—TO ESTABLISH.

exit   An actor's cue to leave the stage or camera's frame.

EXT   Abbreviation for *exterior.* Used in scene headings.

extreme close-up (ECU)   An extremely close camera focal length. In an
EXTREME CLOSE-UP of a face, we would see only a part of the
entire face. The lips, for instance, or eyes might fill the entire
screen.

fade in/fade out   Editing term used to begin and end a script or note a major transition. FADE means the picture fades in from black or some other color or fades out to that color.

FG   Script abbreviation meaning *foreground.*

flat fee   A payment arrangement in which the writer is offered a preset amount of money regardless of the page count or running time of the script.

freeze frame   A picture which is stopped at some key spot often for dramatic effect, such as FREEZE FRAME of an athlete crossing the finish line in a race. Also used to create a still image background panel for superimposed titles.

high angle   A camera angle shot from above, looking *down* on the action.

host   An on-camera spokesperson who talks directly to the audience (camera).

insert   Usually a CLOSE-UP or EXTREME CLOSE-UP of an object, which is inserted into the main action. As an example, a character may begin dialing the telephone in a WIDE SHOT. An INSERT of her hand turning the dial is cut in so we are able to recognize the number she is calling. When the dialing is completed and the woman begins to talk, we return to the WIDE SHOT or some other angle.

instructional design   A method of custom designing instructional material based on criteria such as audience factors, objectives, the problem to be solved, and program utilization.

INT   Abbreviation for *interior.* Used in scene headings.

key   Short for *Chromakey.* It is used to denote the combining of pictures or a picture and title, for example, KEY IN LOGO OVER SHOT or KEY TITLES. It is sometimes used synonymously with SUPER, meaning superimpose.

limbo   A seamless, usually featureless background, often of one color, used as a generic set environment. A spokesperson or a vignette using role-play actors might be set in limbo. This is usually for a dramatic effect or because it is cheaper than building a set.

low angle   A camera angle shot from below, looking *up* at the action.

master scene   A complete scene described in terms of action rather than camera angles and focal lengths. Master scenes are highly descriptive but leave the camera positions and focal lengths to the director and the director of photography. Most directors and producers prefer working with scripts written in master scenes.

master script   The script brought to the field on which the production notes are taken during the shoot. Later it is used by the editor to locate the proper take. It is often filed and retained when the program is completed.

medium close-up (MCU) A camera focal length that frames an individual's face and upper body starting from the chest.

medium shot (MS) A camera focal length in which the majority of a person's body can be seen. A MEDIUM SHOT is framed roughly from the thighs up; a MEDIUM TWO SHOT would include two people in about the same focal length.

montage A series of images, often fast paced and cut to music, used to suggest a compression of time. It is also used for dramatic effect, many times in the openings of programs.

music up/music under/music out Script notations denoting a music increase in volume (MUSIC UP), decrease to allow other sounds more prominence (MUSIC UNDER), or fade out completely (MUSIC OUT).

narration Words spoken by a narrator or host. Narration is spoken directly to the audience (camera).

narrator An off-camera spokesperson. A host is an on-camera version of a narrator.

OC or OS OFF-CAMERA or OFF-SCENE. It means generally the same thing as VO (voice-over). A voice heard while the speaker is not seen, it is usually used in the dialogue or narration heading.

over the shoulder (OTS) A camera angle looking over one person's shoulder at another person. OTS—JILL means we are looking at Jill over someone else's shoulder.

pan Short for *panorama*. A horizontal rotation of the camera on its head.

ped (up or down) Short for *pedestal*. Raising or lowering of the camera on its head.

politics (corporate) Personal preferences and maneuvers that affect what should be strictly business decisions. For instance, a finance manager might not like a peer in the accounting department; therefore, he rejects a valid proposal (which could be script related) that she presents.

POV Script abbreviation meaning *point of view*.

producer The person responsible and accountable for all aspects of a film, video, or audio production from script development through shooting and editing.

program design The process of custom designing a film or videotape production based on factors such as the problem, an audience analysis, specific objectives, and utilization.

program-needs analysis (PNA) A short factual analysis used as a program design document and a justification for producing the program. It is usually developed just after the initial script meeting.

rack focus A change of focus to create emphasis or change the viewer's perspective. As an example, the camera might rack focus from a street sign in the foreground to a man walking in the distance.

reveal   A ZOOM, PAN, or other camera move that reveals something the viewer had not previously been aware of, for example, ZOOM OUT TO REVEAL John entering the office.

reverse angle   A view that is 180 degrees from the last shot. As an example, if we are looking over the shoulder of a driver moving down the street in a car, a REVERSE ANGLE would be the view looking directly backward out the rear window.

rewrite   A major overhaul of an existing script.

role-play   A scene designed to dramatize a real-life situation. It can be serious or humorous.

screenplay format   Script format in which the scene descriptions are written across the entire page and the sound is written in a column down the center. It is the standard "Hollywood" format.

screen right/left   The position of something in the scene as viewed from the *camera's* perspective. The direction is the opposite from anyone on the set who is looking back at the camera. An actor's right, when looking at the camera, is screen left.

script   The framework for a film, television, or audio production. The script is a precisely written visual description of all program elements, which is used by the director and producer to then create the program on film or tape. It is also a product for sale.

shooting script   The final draft with all changes made. A production crew takes the shooting script to the field.

shock cut   A cut used to dramatize a transition and shock the audience. As an example, as a gun is about to be fired at the protagonist, a script might then call for a SHOCK CUT to an EXTREME CLOSE-UP of a telephone ringing in his wife's office.

single   One person only; usually seen in a MEDIUM CLOSE-UP or MEDIUM SHOT.

SME   Subject matter expert (see Content Expert).

SOT   Abbreviation for *sound on tape*. It is used to indicate that sound is coming from some prerecorded tape source, rather than live from an actor or narrator.

sound effect   Notation written into a script to call for a sound effect. It is also called or written in as SFX.

stage (up/down)   Upstage is the area in the background, farthest from the camera; downstage is in the foreground, closest to the camera.

sting   A short piece of music used to punctuate a transition or heighten a dramatic moment.

stock footage   Footage pulled from a library or archive to be included in a program.

story outline   Another term for treatment. A scene-by-scene description of the story written in simple narrative terms.

super   Short for *superimposition*. It means something superimposed over a picture. An example would be SUPER OPENING TITLES.

talent   Another word for *actors*.

technical advisor   A subject matter expert who is usually present on the set during shooting. He helps assure that the scene has been recorded accurately.

tight shot   Using a tight camera frame, such as in TIGHT TWO SHOT.

tilt (up or down)   A vertical rotation of the camera on its head.

treatment   A scene-by-scene narrative describing the story and visual aspects of a script that is yet to be written. Used as a pre-script-approval step to be sure clients and producers like the concept the writer has come up with.

truck   A movement of the camera in a sideways motion, right or left. This is opposed to a DOLLY, which is a movement forward or backward. DOLLY is often used interchangeably with TRUCK.

two shot   A shot of two actors, positioned close together, usually in conversation.

video   The picture column in a two-column format script. It also refers to the video portions of recorded programs and is used loosely to refer to entire programs and the entire field.

vignette   See Role-play.

voice-over (VO)   Portions of narration or dialogue heard while the viewer is seeing something other than the person speaking. It means a voice heard over pictures.

wide shot (WS)   A camera focal length encompassing the entire scene or a large part of it. It is often used to establish a room or environment in which the scene is taking place.

wipe   A transitional effect in which one picture is wiped off the screen by another. It can take place horizontally, vertically, or in circular or rectangular patterns.

zoom in/zoom out   A change of focal length by the camera. As an example, from a WIDE SHOT, the script might call for a ZOOM IN TO CU—JENNY.

Bibliography

De Luca, Stuart M., *Instructional Video*. Boston: Focal Press, 1991.

DiZazzo, Ray. *Corporate Television*. Boston: Focal Press, 1989.

Gayeski, Diane M. *Corporate & Instructional Video Design and Production*. Englewood Cliffs, NJ: Prentice Hall, 1983.

Matrazzo, Donna. *The Corporate Scriptwriting Book*. Portland, OR: Communicom, 1985.

Swain, Dwight V. and Joye R. Swain. *Scripting for the New AV Technologies*, 2nd ed. Boston: Focal Press, 1991.

Swain, Dwight V., with Joye R. Swain. *Film Scriptwriting: A Practical Manual*, 2nd ed. Boston: Focal Press, 1988.

Van Nostran, William J. *The Scriptwriter's Handbook*. White Plains, NY: Knowledge Industry Publications, 1989.

INDEX